INSTITUTE OF GEOLOGICAL SCIENCES
Natural Environment Research Council

British Regional Geology
# The Hampshire Basin and adjoining areas

FOURTH EDITION

By R. V. Melville, MSc and
E. C. Freshney, BSc, PhD

LONDON   HER MAJESTY'S STATIONERY OFFICE   1982

*The Institute of Geological Sciences was formed by the incorporation of the Geological Survey of Great Britain and the Museum of Practical Geology with Overseas Geological Surveys and is a constituent body of the Natural Environment Research Council*

*© Crown copyright 1982*
*First published 1936*
*Fourth edition 1982*

ISBN O 11 884203 X

# Foreword to Fourth Edition

When, in 1935, the Museum of Practical Geology was transferred from Jermyn Street, Piccadilly, to Exhibition Road, South Kensington, the purpose of illustrating the geology of Great Britain was served by dividing the country into eighteen regions. To each region a bay in the Museum was allotted for the display of specimens, photographs and maps. At the same time a set of short, illustrated handbooks descriptive of the geology and scenery of the different regions was instituted, to serve as guides to the Museum exhibits and also to provide concise explanations of the geology for use in the field.

The first edition of the Hampshire Basin handbook was written by the late Mr C. P. Chatwin and published in 1936. It was followed by second and third editions in 1948 and 1960. With the passage of time knowledge has both increased and diversified, while interpretations have changed, so for this fourth edition the text has been revised and in large part rewritten by Mr R. V. Melville. The chapter on Structure has been contributed by Dr E. C. Freshney and includes a contour map of the base of the Tertiary which was prepared by Mrs R. Moseley. The book has been edited by Mr G. Bisson.

The authors wish to acknowledge help from many sources, particularly Dr R. G. Clements, Dr B. W. Sellwood, Dr W. G. Townson and Dr I. M. West on Jurassic geology; and Professor D. Curry, Dr C. Downie, Professor J. W. Murray and Mr F. C. Stinton on Tertiary geology. Their colleagues in the Institute of Geological Sciences have been unfailingly helpful.

Institute of Geological Sciences  
Exhibition Road  
London SW7 2DE  
26 February 1982

G. M. BROWN  
*Director*

*An EXHIBIT illustrating the geology and scenery of the region described in this handbook is set out in the Geological Museum, Institute of Geological Sciences, Exhibition Road, South Kensington, London, SW7 2DE.*

# Contents

|   | | Page |
|---|---|---|
| 1 | **Introduction** | 1 |
| 2 | **The geological framework** | 2 |
| 3 | **Jurassic rocks:** Introduction; Lias or Lower Jurassic (Lower Lias, Middle Lias, Upper Lias); Middle Jurassic (Inferior Oolite, Great Oolite Series Dorset Basin, Great Oolite Series northwards from Westbury); Upper Jurassic (Kellaways Beds and Oxford Clay, Corallian Beds, Kimmeridge Clay, Portland Group, Lower part of Purbeck Group, Lulworth Formation) | 7 |
| 4 | **Cretaceous rocks:** Purbeck Group of Cretaceous age (= Durlston Formation) and Wealden; Lower Greensand; Gault, Upper Greensand and basal beds of the Chalk; Chalk (Lower Chalk, Middle Chalk, Upper Chalk) | 62 |
| 5 | **Tertiary (Palaeogene) rocks:** Introduction; Palaeocene (Reading Beds, London Clay Basement Bed); Eocene (London Clay above Basement Bed, Bagshot Beds of Isle of Wight, Bracklesham Group, Barton Formation, Solent and Bembridge formations); Oligocene (Hamstead Formation) | 90 |
| 6 | **Structure:** Introduction; Structures trending approximately east–west; Structures trending approximately north-west–south-east; Influence of structure on sedimentation; Structural history of the Hampshire Basin | 112 |
| 7 | **Pleistocene and Recent:** Introduction; Clay-with-flints; Plateau Gravels, Valley Gravels and Alluvium; Brickearth, Coombe Rock and Head; Raised Beaches and Submerged Forests; The Chesil Bank and The Fleet; History of the Pleistocene Period | 124 |
| 8 | **Selected references** | 132 |
|   | **Index** | 139 |

## Illustrations

**Figures** **Page**

1. Geological sketch-map of the Hampshire Basin and adjoining areas — Facing 1
2. Section along the cliffs from Pinhay Bay past Lyme Regis and Charmouth to near Seatown — 11
3. Section along the cliffs from Seatown to Bridport Harbour and Burton Bradstock — 12
4. Outcrop of the Yeovil and Bridport Sands showing the estimated position of the bar front and tidal channel at the time of formation of the Ham Hill Stone — 19
5. Vertical section through the Inferior Oolite at Burton Bradstock — 22
6. Vertical section through the Inferior Oolite at Halfway House — 23
7. Variations in thickness and facies in the Great Oolite Series between Bath and Lyme Bay — 24
8. Section along the cliffs from south of Weymouth north-east to Redcliff Point — 35
9. Vertical section through the Corallian Beds east of Osmington Mills — 36
10. Trigonia clavellata Beds, Osmington Mills — 37
11. Section along the cliffs near Kimmeridge Bay, Dorset — 41
12. Comparative classifications of the Portland Group in Dorset — 44
13. Correlations and facies variations in the Portland Group in Dorset — 46
14. General model for environments of deposition of the Portland Group of Dorset — 54
15. General model of maximum observed diversity of environments of deposition of the Portland Group of Dorset — 55
16. Section from Durlston Head northwards to Peveril Point, Swanage — 57
17. Stratigraphy and lithology of the basal Purbeck Group (Lulworth Formation) of the Dorset mainland — 58
18. Sections of the Wealden Beds in the Isle of Wight — 66
19. Cliff section of Lower Greensand, from Atherfield Point to St Catherine's Hill, Isle of Wight — 68
20. Relationships of Albian and Cenomanian strata in Dorset and the Isle of Wight — 75–76
21. Diagram of Cenomanian transgression across Dorset — 78
22. Upper Greensand and Lower Chalk in two sections near Warminster — 82
23. Generalised succession of facies in the Palaeogene beds of the Hampshire Basin — 91
24. Section from Melcombe Park to near Wool, Dorset — 92
25. Section of Eocene beds in Alum Bay, Isle of Wight — 100
26. Foreshore exposures of Bracklesham Group and Quaternary deposits between Bracklesham and Selsey Bill, Sussex — 101
27. Facies and correlation in the Bracklesham Group — 103

| Figures | | Page |
|---|---|---|
| 28 | Cliff sections between Poole Harbour and Boscombe (now largely obscured) through Bournemouth Freshwater Beds and Bournemouth Marine Beds | 104 |
| 29 | The Barton Cliff section | 106 |
| 30 | Contours on the base of the Tertiary | 113 |
| 31 | Principal surface structures in the Hampshire Basin region and postulated basement fault blocks | 115–116 |
| 32 | Interpretation of the structures associated with the Chaldon Pericline | 117 |
| 33 | Major structural elements of the Weymouth – Swanage area | 118–120 |
| 34 | Relationships of present drainage to conjectured Plio-Pleistocene marine transgression | 119–120 |
| 35 | Hypothetical reconstruction of the 'Solent river system' at a time of low sea level | 131 |

| Tables | | |
|---|---|---|
| 1 | Main stratigraphic units recognised in the Hampshire Basin and adjoining areas | 4–5 |
| 2 | Jurassic succession on the Dorset coast | 9 |
| 3 | Biostratigraphy and lithostratigraphy of the Lias of the Dorset coast | 13 |
| 4 | Stratigraphy of the Kellaways Beds and Oxford Clay | 31 |
| 5 | Biostratigraphy and lithostratigraphy of the Kimmeridge Clay of Dorset | 42 |
| 6 | Thickness of the Lulworth Formation in Durlston Bay | 56 |
| 7 | Cretaceous succession in the Isle of Wight and Dorset | 63 |
| 8 | Thickness of the Durlston Formation in Durlston Bay | 64 |
| 9 | Stratigraphy of the Lower Greensand at Atherfield, Isle of Wight | 69 |
| 10 | Stratigraphic classification of the Chalk | 84 |
| 11 | Classification of the Palaeogene Beds of the Hampshire Basin | 98–99 |
| 12 | Stratigraphy of the Barton Formation | 107 |
| 13 | Stratigraphy and suggested chronology of the Pleistocene in the Hampshire Basin | 129 |

**Front cover** St Alban's Head from the air, showing the Portland Limestone Formation overlying the Portland Sand Formation in the undercliff.

| Plates | | Page |
|---|---|---|
| 1 | Fossils from the Lower and Middle Lias | 15 |
| 2 | Fossils from the Upper Lias, Inferior Oolite, Fuller's Earth and Bradford Clay | 21 |
| 3 | Fossils from the Cornbrash and Oxford Clay | 29 |
| 4.1 | Cliffs of Upper Lias, Bridport Sands, West Bay, showing bands of calcareous sandstone weathered out by wind-blown sand. (A.12038) | |
| 2 | Cliffs of Lower Lias, Blue Lias, east of Lyme Regis, showing alternation of limestone and shale. (A.12036) | 33 |

| **Plates** | **Page** |
|---|---|
| 5.1 Quarry face in Great Oolite limestones showing Twinhoe Beds (well-bedded limestone) resting on Combe Down Oolite. Winsley Quarry, near Bath. (A.9734) | |
| 2 Cliffs west of Kimmeridge Bay showing Kimmeridge Clay with, in foreground, the hard ledge of Broad Bench. (A.12230) | 34 |
| 6 Fossils from the Corallian Beds and Kimmeridge Clay | 45 |
| 7 Fossils from the Portland and Purbeck Groups | 47 |
| 8.1 Cliffs south of Blacknor, Isle of Portland, showing beds of Portland Group capped by Lulworth Formation. (A.12236) | |
| 2 Coastal landslip of the Lower Greensand in cliffs north-west of Blackgang Chine, Isle of Wight. (A.12024) | 51 |
| 9.1 Stair Hole, west of Lulworth Cove, showing beds of Lulworth Formation dipping steeply to the north in the middle limb of the Purbeck monocline, and folded in the 'Lulworth Crumple'. (A.11223) | 52 |
| 2 Fossil forest, east of Lulworth Cove. (A.12224) | |
| 10 Some characteristic ostracods from the Purbeck Group | 59 |
| 11 Fossils from the Wealden Group and the Lower Greensand | 73 |
| 12 Fossils from the Gault, Upper Greensand and Lower Chalk | 81 |
| 13.1 Cliffs of Durdle Cove, Dorset, showing a vertical bed of Chalk eroded by wave action along a thrust-plane. (A.12229) | |
| 2 Dry valley in Chalk country near Warminster, Wiltshire. (A.9708) | 85 |
| 14 Fossils from the Chalk | 87 |
| 15 Fossils from the Eocene | 97 |
| 16 Fossils from the Eocene and Oligocene | 111 |
| 17.1 Chesil Beach, from West Cliff, Isle of Portland. (A.12233) | |
| 2 Trail of sarsen stones brought to their present position by solifluction, Fyfield, near Marlborough. (A.11411) | 127 |

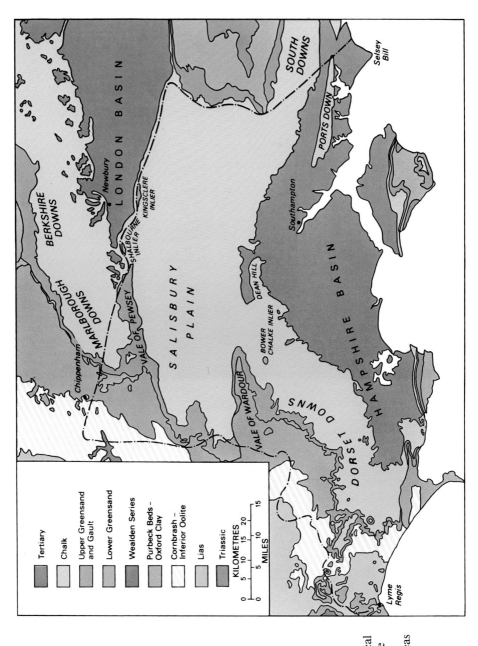

**Figure 1** Geological sketch-map of the Hampshire Basin and adjoining areas

# 1. Introduction

The region described in this Handbook—the Hampshire Basin and adjoining areas—includes the whole of Dorset, the greater part of Wiltshire and Hampshire, the Isle of Wight and part of West Sussex (Figure 1). In this region, formations of the Jurassic, Cretaceous, Eocene and Oligocene systems are exposed, as well as a variety of superficial deposits. The Handbook is prepared for the reader who is at an early stage of formal study, or who has a general interest in the region; more detailed and technical accounts are included in the Memoirs of the Geological Survey and in the general geological literature. A list of selected references is provided at the end of the book.

The accessibility of the coast sections of Dorset and the Isle of Wight, and the richly fossiliferous nature of many of the formations exposed, have made the reputation of the region among professional and amateur geologists for nearly 200 years. A book longer than this one could easily be written on the contributions that the area has made to stratigraphy, palaeontology and structural geology in both their theoretical and practical aspects, as well as to more modern studies in sedimentology and ichnology (trace-fossils). The references have been chosen so as to include a number of works helpful to those interested in following up these matters.

In all Britain, no area is under such heavy pressure as this from visits by school and university parties and from the activities of those who collect fossils for pleasure or profit. Visitors to the coast sections should have due regard to their own safety where tides and cliff-falls are concerned. They should also remember that these sections are classic documents in the history of geological science world-wide and, as such, are visited by geologists from many countries. Their scientific value should therefore be respected, and not lessened by wanton hammering.

Inland sections should not be visited without the prior consent of the landowner or responsible authority. Lumps of rock should not be left where they can damage machinery or injure animals or people.

Fossils, rocks and minerals may be attractive ornaments; they are also a store of scientific—geological and biological—information which is wasted if the greatest possible amount of detail of locality and horizon is not recorded when they are collected. Every rock that is split reveals something never before seen by the eye of man; and every new specimen may have something to teach the most learned among us.

# 2. The geological framework

The visible story of the geology of the region begins with the dawn of the Jurassic Period, represented by the basal Liassic deposits west of Lyme Regis (Table 1). Little is certainly known of earlier events: it is, for instance, not known where the Carboniferous rocks pass from the Culm facies of Cornwall and Devon to the Carboniferous Limestone – Millstone Grit – Coal Measures facies exposed in the Bristol and Gloucester district and proved beneath the Weald of Kent. This transition is buried beneath the area. It is thought that some of the major structures that we see (the Purbeck and Weymouth anticlines and some of the major faults) are related to the revival of Variscan structures of east – west trend in buried rocks. It is also uncertain how far the Triassic rocks, known to contain salt deposits in north-west Somerset, continue to do so under this area; though a positive gravity anomaly near Fordingbridge and the peculiar Compton Valence anticline, formerly attributed to the displacement of strata by the plastic flow of salt under pressure at depth, now seem to reflect basement rocks at shallow depth.

During most of the Jurassic the area lay at about 30°N latitude—the parallel of present-day Cairo and Abadan—and formed part of a shallow sea separating land masses to east and west. According to Casey (1971), by the end of Kimmeridgian times Britain lay on the eastern flank of a typical cyclonic, oceanic circulatory system—the ocean being the north Atlantic, transformed by the westward drift of the American continental plate from a shelf sea to a true ocean. Water welling up from the depths in such conditions would be rich in nutrients, and the idea is supported by the fact that episodes of inhibited deposition from that time on are distinguished from earlier ones by the abundance of associated phosphate. Even so, this sea was of very wide extent (though nearly landlocked in the area of the British Isles) and for most of the time very shallow. Relatively few beds show certain evidence of having been formed below the photic zone, which is the depth (about 80 m) reached by daylight; and some formations were laid down between tide-marks or even above the level of normal tides. To the south, across present-day central and southern Europe and the Mediterranean, lay the wide and deep Tethyan ocean.

The depth of the sea fluctuated in response to world-wide (eustatic) movements and smaller-scale but still continent-wide (epeirogenic) movements, of amplitudes generally less than 30 m and with little, if any, horizontal component. Such fluctuations would produce no perceptible effect on oceanic deposits, but in shallow seas of wide extent pronounced changes would occur. For example, the thick clay formations of the Jurassic may have been laid down in water of practically the same, if moderate, depth and sluggish circulation (so far as we can now tell) over the whole area; but a regional shallowing would soon enhance the importance of quite small depth differences. Some areas might emerge above the sea, or become too shallow to receive sediment. Others might be isolated by offshore barriers which would separate areas of different facies.

Even during periods when rocks of practically uniform facies were being deposited in water of practically uniform depth, the rate of subsidence may not have been uniform—and subsidence had to continue to allow sediments to accumulate. Broadly speaking, the area of southern Dorset, southern Hampshire and the Isle of Wight behaved as a group of subsiding basins throughout Mesozoic and early Tertiary times. Kent (1949) estimated the maximum depth to the base of the Mesozoic in this area at 3658 m. This group of basins was bounded by a north-westerly trending structural line—a continuation of the Bembridge–St Valéry line (see Chapter 6)—which marked the boundary of the Wealden Basin to the north-east (outside the area).

In early and middle Jurassic times there is no direct evidence of where the western margin of the basin of deposition lay. In late Kimmeridgian times, however, lithological changes suggest that it lay not far west of the present outcrop. From Volgian to middle Albian times, the margin may have lain within the region, while from late Albian to Turonian times, thinning and facies-changes in the Upper Greensand and Chalk indicate a marked local shallowing of the sea. In early Tertiary times there is no reason to suppose that the basin of deposition extended far beyond the present westerly outcrop of beds of that age. These phenomena may indicate upward movement in deeply buried basement rocks from late Jurassic times onwards, including upheaval of the ancient granite mass that extends from Dartmoor to the Isles of Scilly.

The present disposition of the rocks is due in large part to such differences as these in the rate of subsidence between graben and horst portions of the area (see Chapter 6). Allowance must also be made, however, not only for variations in thickness due to the same cause, but also for the powerful effects of the tectonic earth movements which affected southern Britain in two main spasms—in early Cretaceous (pre-Albian and pre-Gault) times, and in Miocene (probably late Miocene) times; the movements that are expressed by the unconformity between the Chalk and the overlying Eocene were less dramatic. Thus the base of the Lower Lias, which is at sea level west of Lyme Regis, is at 1432 m below OD at Bere Regis and at 1903 m below OD at Portsdown. The base of the Cornbrash is more than 122 m above OD WNW of Chippenham, at 713.8 m below it at Bere Regis, and at 1368 m below OD at Arreton (Isle of Wight). The base of the Chalk, which rises above 152 m near Buttermere, on the northern edge of the region, is at 539 m below sea level at Fordingbridge, and is probably deeper than this in the axis of the Isle of Wight syncline. Tectonic movements have, of course, raised levels in anticlinal areas and depressed them in synclines; no attempt has been made to discount these effects so as to calculate the potency of epeirogenic and eustatic subsidences as such.

During at least parts of Jurassic and late Cretaceous times, land lay far enough to the west and south-west of the region for possibly the whole of Wales to be under the sea of the continental shelf. In late Jurassic and early Cretaceous times, however, those areas and probably also the western and north-western parts of the region, were land, fringing first an area of lagoons and sand bars with braided channels, and then a shallow sea, in which the Lower Greensand was deposited.

In Tertiary times, the sea was confined to a comparatively small area—the Hampshire Basin strictly defined (though with a connection to the London

# 4  The Hampshire Basin

**Table 1** Main stratigraphic units recognised in the Hampshire Basin and adjoining areas (not to scale). Stages in parentheses are not represented by marine deposits in this region

| Estimated Ages Ma | Chronostratigraphic Periods | Stages | Lithostratigraphic Formations | Systems |
|---|---|---|---|---|
|  | Recent |  | Blown sand, Shingle beaches, Alluvium, Tufa, Peat | Recent |
|  | Pleistocene |  | Raised beach, River terraces (gravels), Coombe deposit, Angular flint gravel, Plateau gravel, Brickearth and Clay-with-flints | Pleistocene |
| 38 | Oligocene | Rupelian | Hamstead | Oligocene |
|  | Eocene | Bartonian Lattorfian | Bembridge | Eocene |
|  |  |  | Solent |  |
|  |  | Ludian | Barton |  |
|  |  | Marinesian |  |  |
|  |  | Auversian | Bracklesham Group  Selsey |  |
|  |  | Lutetian | Earnley |  |
|  |  | Cuisian | Wittering |  |
|  |  | Ypresian | London Clay |  |
| c.64 | Palaeocene |  |  | Palaeocene |
|  |  | Sparnacian | Reading Beds |  |
|  | Upper Cretaceous | Campanian | Upper Chalk | Upper Cretaceous |
|  |  | Santonian |  |  |
|  |  | Coniacian |  |  |
|  |  | Turonian | Middle Chalk |  |
| c.100 |  | Cenomanian | Lower Chalk |  |

*Geological Framework* 5

| Estimated Ages Ma | Chronostratigraphic | | Lithostratigraphic | |
|---|---|---|---|---|
| | Periods | Stages | Formations | Systems |
| | Lower Cretaceous | Albian | Upper Greensand and Gault | Lower Cretaceous |
| | | | Lower Greensand | |
| | | Aptian | | |
| | | (Barremian) | Wealden | |
| | | (Hauterivian) | | |
| | | (Valanginian) | — — — — — — — — — — | |
| 135 | | (Ryazanian) | Purbeck Group   Durlston Formation | |
| | Upper Jurassic | Port-landian | Lulworth Formation | Upper Jurassic |
| | | Volgian | Portland Group | |
| | | Kimmer-idgian | Kimmeridge Clay | |
| | | Oxfordian | Corallian Beds | |
| | | Callovian | Oxford Clay and Kellaways Beds | |
| | Middle Jurassic | Bathonian | Great Oolite Series | Middle Jurassic |
| | | Bajocian | Inferior Oolite | |
| | Lower Jurassic | Toarcian | Upper Lias | Lower Jurassic |
| | | Pliensbachian | Middle Lias | |
| | | | Lower Lias | |
| | | Sinemurian | | |
| 200 | | Hettangian | | |

*Note* Since this guide was written modern international usage and the classification in the Geological Society's Special Report on Jurassic correlation, have placed the Callovian in the Middle Jurassic.

Basin) and the Isle of Wight. Even within these limits, the sea was not continuous, and the latest pre-Pleistocene deposits now preserved—the Hamstead Beds—are in large part of freshwater origin. If later Oligocene and perhaps Miocene deposits were laid down, no trace of them now remains. In Miocene times the area was uplifted by the Alpine mountain-building movements, since when thousands of metres of material have been removed by erosion. In the Pleistocene, though the English Channel may for a time have held a lobe of ice extending from the west past Selsey Bill, there is no evidence that the whole area was glaciated. After the climax of the last (Devensian) glaciation, immense volumes of water drained off the higher ground into a river which followed the line of the Solent, leaving behind an intricate series of gravels.

The present-day scenery of the region is dominated by the Chalk uplands of Hampshire—with Inkpen Beacon, at just over 305 m the highest Chalk hill in England—and Salisbury Plain. From here a south-westerly lobe of Chalk—the Dorset Downs—stretches across mid-Dorset to beyond the region. South and west of this lobe lies the tumbled Jurassic country of west Dorset and the hinterland of the Chesil Beach. The Chalk reaches the sea east of Weymouth, cutting off the Isle of Purbeck from the heathland around Wareham. North of the Dorset Downs the Jurassic clay country of Blackmore Vale reaches far into the Chalk country along the anticlinal axis of the Vale of Wardour—nearly matched by the Vale of Pewsey farther north. North-west from here lies the dip-slope of Jurassic clays rising to the limestone scarp of the south Cotswolds.

From Salisbury Plain the famous trout streams of Hampshire drain down through the soft-bottomed ground of the New Forest to Southampton Water and the sea. To the east lie the drowned marshes of Langston and Chichester Harbours. The Isle of Wight was probably cut off from the mainland only during the last Pleistocene glaciation, when the Chalk ridge between The Needles and Handfast Point was breached by the combined action of the sea and the Solent river. Thus Poole Bay and Christchurch Bay may be no more than some 15 000 years old.

The excellent exposures of fossiliferous strata along the coast in this region allow the two autonomous methods of stratigraphical classification (lithostratigraphy and chronostratigraphy) to be clearly understood. Lithostratigraphy recognises formations as masses of rock distinguished by their physical and mineralogical characters and by their position in the sequence. Chronostratigraphy recognises stages and zones bounded by presumed time-planes which (so far as this area is concerned) are represented either by non-sequences or by changes in the fossils (i.e. by biostratigraphic evidence). The planes separating the units of one kind in a given area need not always coincide with those separating units of the other kind. This is brought out for the stages and 'formations'[1] in Table 1, and is shown in more detail in later tables.

---

[1] This term is used in a general sense, without prejudice to formal classification of lithostratigraphic units in the hierarchy Group, Formation, Member, Bed.

# 3. Jurassic rocks

## Introduction

The whole of the Jurassic System is exposed along the Dorset coast between Lyme Regis and Swanage (Table 2). There are only a few small time-breaks in the succession, which is about 1350 m thick. The whole is marine, except for the Lulworth Formation of the Purbeck Group at the top which was laid down partly in fresh, partly in hypersaline water. The rocks are less well known inland because, even where not overlapped by Cretaceous rocks, they are seldom exposed apart from small quarries and temporary sections.

The succession comprises a number of major clay–silt–sand–limestone sequences, each of which represents deposition in a shallowing sea. Each of these major sequences is composed of many smaller-scale units made up in the same way. Each rhythmic unit, or cyclothem, begins with a clay, which is usually the thickest member. The clay may be a fissile shale (when it may also be bituminous) or a more massive mudstone, pyritic and with or without calcareous nodules. When fossils occur their original skeletal structures may be preserved in fine detail. A clay that is not fossiliferous, especially if it lacks signs of the churning and burrowing action of animals living on or within the sea floor—'trace-fossils'—is taken to have been formed in water without dissolved oxygen, and generally speaking in comparatively deep water. As a general rule, the deposition of a fossiliferous clay would have been faster than the rate of subsidence: clay would then have accumulated until the sea floor had been raised within the reach of wave action (Sellwood and Jenkyns, 1975); clay would then have been winnowed away and silt deposited.

The next member of a cyclothem is a sand—bright yellow at outcrop, but usually bluish grey at depth, and in many cases not sharply separated from the clay beneath. Some sands contain more or less regular layers of calcareous sandstone; others contain spheroidal calcareous nodules ('doggers') up to 5 m or more in diameter. Fossils, where preserved, tend to be found in nests or clusters and to be heavy-shelled types. Most sands show abundant trace-fossils and signs (such as ripple-marks) of having been deposited in moving water.

The top unit is a limestone, in many cases containing ferruginous grains of one form or another. Fine-grained limestones (as in the Lias) mark virtual cessation of deposition and may show signs of active erosion; for this reason they are taken to mark the shallowest phase of each cycle, though Sellwood and Jenkyns (1975) have shown that this is an intuitive, not an inductive conclusion. Shelly limestones, and particularly oolites, were probably deposited very rapidly and in extremely shallow, at times supra-tidal, water. The clay of the succeeding cyclothem follows abruptly.

Most of the Jurassic rocks of the area show abundant trace-fossils. The four commonest types are:

*Diplocraterion* U-shaped burrows with reworked sediment between the arms of the U. The burrows are at right angles to the bedding. The whole measures up to

21 x 4 cm; each arm of the tube is up to 0.7 cm in diameter. These were the permanent domicile of worms or crustaceans that angled or swept for food suspended in the water. Common in shallow marine or estuarine beds deposited quickly, e.g. Bencliff Grit.

*Chondrites* Regularly branching, vertical to horizontal burrows spreading down like the roots of a tree, and 2 mm or more in diameter. The work of sediment-eating animals, perhaps sipunculoid worms.

*Thalassinoides* Ramifying, Y-branching networks of plain horizontal and vertical tubes, 1 to 5 cm in diameter. Feeding burrows of decapod crustaceans.

*Rhizocorallium* Horizontal to oblique U-tubes, each arm 1 cm or more in diameter; the arms several centimetres apart with reworked sediment between them. Tubes short or long depending on whether the animal was in a suspension-feeding or a deposit-feeding phase of activity.

Sellwood (1970; see also Sellwood and others, 1970) has shown how trace-fossils demonstrate the primary difference (which may have been secondarily enhanced) between the shale and limestone members of small-scale Lower Lias cyclothems. These usually begin with bituminous shales formed in practically anaerobic conditions. The succeeding mudstones or marls accumulated in aerated water and were thoroughly bioturbated by *Chondrites*. The tops of the beds are usually sharply defined and, like the succeeding limestone, suggest formation in turbulent conditions. Sediment-eaters burrowed in all directions and *Rhizocorallium* is common. There was, in fact, a progressive increase in the diversity of the fauna (presumed to be crustaceans and worms) responsible for the trace-fossils as each cycle proceeded. The progressive physical changes already described (p.7) would have enabled the sea-floor to support more life and would have brought it nearer to the steady rain of dead plankton falling from the surface layers of the sea to nourish the animals living on and within the bottom.

Sellwood and Jenkyns have suggested modifications of earlier authors' views about 'basins' (with thick sedimentary sequences) and 'swells' (with thin sequences) in the Jurassic, for they showed that swells were both local (passing laterally into basins) and impermanent (behaving as basins at certain times). It follows that episodes of swell or basin deposition were not synchronous over the whole of the British area. Hallam and Sellwood (1976) concluded that the pattern of sedimentation was controlled by two major influences: (a) tension giving rise to vertical movement along faults in the basement rocks, and (b) regional upwarping or doming of the crust, followed by collapse of the domed area in late Jurassic times. The former, which was intermittently active, would explain the behaviour of basins and swells. The latter would account for the steady rise in sea level which is apparent throughout the Jurassic of north-west Europe, followed by the development of a relatively deep and wide basin in Kimmeridgian times.

## Lias or Lower Jurassic

The oldest rocks that crop out in this region belong to the Lias. This formation (its name probably came from the Gaelic word *leac,* a flat stone, see Plate 4.2) consists largely of clays and limestones but includes also beds of shale, layers of calcareous nodules, and beds of sand and sandstone. Fossils are common and show that nearly all the beds were deposited in the photic zone; remains of saurian reptiles and insects indicate that land was not far distant.

**Table 2** Jurassic succession on the Dorset coast

| GEOLOGICAL AGE | | | ENVIRONMENT<br>Non-marine / Marine<br>Shallower ← retreating<br>Deeper → advancing | FORMATIONS | MAIN TYPES OF ROCK | Cumulative thickness in metres | Rate of deposition | Theoretical age in millions of years |
|---|---|---|---|---|---|---|---|---|
| UPPER JURASSIC | Portlandian | Volgian | | LULWORTH FORMATION | Non-marine limestones and clays<br>Evaporites | | Slow to moderately fast | 135 |
| | | | | PORTLAND GROUP | Shallow marine limestones<br>Deeper marine dolomites<br>Deeper marine silts and muds | 100<br>200 | Moderately fast | |
| | Kimmeridgian | | | KIMMERIDGE CLAY | Marine shales (bituminous in some beds) with widely-separated limestone bands | 300<br>400<br>500<br>600 | Average<br>Slow to average | |
| | Oxfordian | | | CORALLIAN BEDS | Cyclic shallow marine limestones, sands and clays | 700 | Average | |
| | Callovian | | | OXFORD CLAY, KELLAWAYS BEDS AND UPPER CORNBRASH | Deeper marine clays and shales (bituminous in lower part)<br>Shallow marine sandy limestone at base | 800<br>900 | Average | |
| MIDDLE JURASSIC | Bathonian | | | GREAT OOLITE SERIES | Shallow limestones and clays<br>Deeper marine clays with a few limestones | | Very slow | c.160 |
| | Bajocian | | | INFERIOR OOLITE | Very shallow marine | 1000 | | 170 |
| LOWER JURASSIC | Toarcian | | | UPPER LIAS | Shallow marine sands on sandy clay<br>Shallow limestone at base | | Slow to average<br>Very slow at base | |
| | Pliensbachian | | | MIDDLE LIAS | Deeper marine clays with shallow sands in middle part | 1100<br>1200 | Moderately fast | |
| | Sinemurian | | | | | | | |
| | Hettangian | | | LOWER LIAS | Moderately deep to deeper marine shales with limestone bands mainly in lower part | 1300 | Slow to average | 190 |

In the area the Lias crops out only in Dorset, over a small tract from Lyme Regis to near Bridport and northwards to near Beaminster, thence westwards to the Vale of Marshwood (a periclinal structure lying mostly on the lower beds) and northwards to the Somerset border. A small outcrop occurs northwest of Sherborne. The cliffs between Lyme Regis and Bridport (Figures 2 and 3) expose some of the finest sections of Lias in England. The thickness of the formation in this part of Dorset is more than 244 m.

For the purposes of broad comparison throughout the country the Lias is considered under Lower, Middle and Upper divisions. Local formations within these are recognised on the Dorset coast and have been mapped in the Vale of Marshwood, but cannot all be recognised farther inland. The beds are also classified in zones characterised by the fossils found in them and the zones are grouped into stages. These subdivisions are shown in Table 3.

Hallam (1961) recognised 11 sedimentary rhythms (cyclothems) in the Lias of Britain (see Table 3), not all equally well marked in Dorset. The most important points to note are (a) the erosion and slight non-sequence at the end of cyclothem V; (b) a period of erosion cutting out the *oxynotum* Zone, and another at the end of cyclothem VI, followed by major transgression at the beginning of VII; (c) the non-sequence in cyclothem VIII, followed by widespread uplift and shallowing of the sea in *spinatum* Zone times; (d) the extraordinary condensed deposit known as the Junction Bed, in which are represented cyclothems IX and X of other areas; and (e) the flooding of the whole of the region with sands migrating from the north in cyclothem XI.

## Lower Lias

Hallam (1964) summarised earlier work on the conditions of deposition of the lowest unit of the Lower Lias, the Blue Lias. This formation is famous for the alternations of mudstones or laminated shales (up to 1.8 m thick) with limestones (from 3 to 30 cm thick), which Hallam suggested might have resulted from climatic oscillations, more limestone being formed in warmer conditions. This primary tendency towards an alternation of shale and limestone was accentuated in the diagenetic phase shortly after deposition by solution and re-precipitation of calcium carbonate. Some limestone bands can be traced over several kilometres.

The fossils of the Blue Lias (Plate 1) include numerous ammonites, the small brachiopod *Calcirhynchia calcaria,* bivalves such as *Plagiostoma giganteum* and oysters of the *Liostrea hisingeri – Gryphaea arcuata* series, fragmentary crinoid stems, saurian bones and pieces of lignite up to 2 m long. Near the top of the division are a Fish Bed and a Saurian Bed, and the upper limit is marked by the Hard Marl or Table Ledge (0.3 m thick) seen at Lyme Regis in West Cliff, above the Esplanade, and in Church Cliffs. It descends to the beach at the foot of Black Ven.

The Shales-with-Beef consist of 21.33 m of alternating shales and mudstones with seams of fibrous calcite ('beef') from 1.5 to 102 mm thick. The seams of beef are cross-fibred with cone-in-cone structure. Fossils, particularly ammonites, are common but not well preserved. *Arnioceras* is the most abundant ammonite genus and ranges up from the beds below; *Coroniceras (Arietites)* and *Cymbites* range into higher beds. The lowest 12 m consist mainly of bluish grey conchoidal marls. The upper 9.3 m consist of brown,

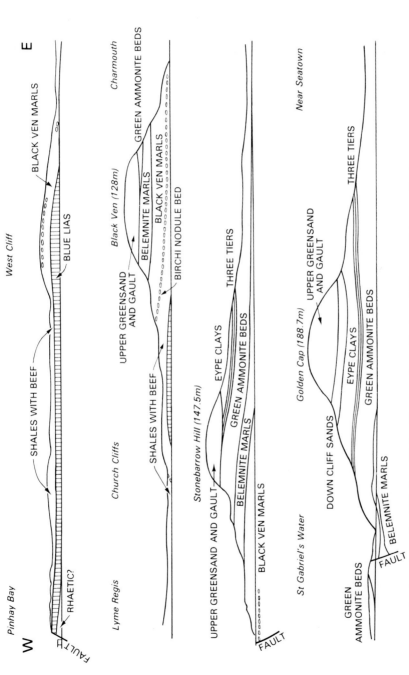

**Figure 2** Section along the cliffs from Pinhay Bay past Lyme Regis and Charmouth to near Seatown. Total distance 9.5 km. Vertical scale exaggerated. (After Woodward, 1893, p.53)

12   The Hampshire Basin

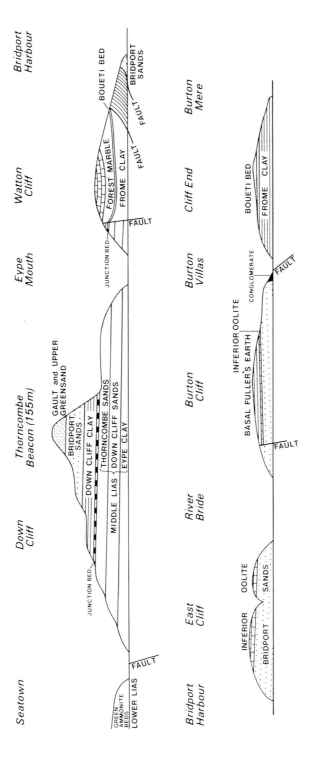

**Figure 3**   Section along the cliffs from Seatown to Bridport Harbour and Burton Bradstock Total distance 9 km. Vertical scale exaggerated. (After Arkell, 1933, p. 154)

**Table 3** Biostratigraphy and lithostratigraphy of the Lias of the Dorset coast

| Stage | Ammonite Zones | Cyclothems of Hallam (1961) | Formations | Lias |
|---|---|---|---|---|
| Toarcian | *Dumortieria levesquei* | XI | Bridport Sands 67·05m | Upper Lias |
| Toarcian | *Grammoceras thouarsense* *Haugia variabilis* *Hildoceras bifrons* *Harpoceras falciferum* *Dactylioceras tenuicostatum* | X  IX | Down Cliff Clay 21·34m | Upper Lias |
| Upper Pliensbachian or Domerian | *Pleuroceras spinatum* | | Junction Bed (incl. Marlstone Rock-Bed) 1·2m | Middle Lias |
| Upper Pliensbachian or Domerian | *Amaltheus margaritatus* | VIII | clays, Thorncombiensis Bed and Thorncombe Sands 26·68m  Margaritatus Clay 2·1m  Margaritatus Stone 0·3m  Down Cliff Sands 26·21m  Starfish Bed  } 2·21m  Day's Shell Bed  Eype Clay 20·25m  Eype Nodule Bed 0·51m  Eype Clays 38·1m  Three Tiers 6·30m   clays 3·35m | Middle Lias |
| Lower Pliensbachian or Carixian | *Prodactylioceras davoei* *Tragophylloceras ibex* *Uptonia jamesoni* | VII  TRANSGRESSION  EROSION | Green Ammonite Beds 32m  Belemnite Marls 22·86m | Lower Lias |
| Upper Sinemurian | *Echioceras raricostatum* *Oxynoticeras oxynotum* *Asteroceras obtusum* | VI EROSION  V | Black Ven Marls 45·72m | Lower Lias |
| Lower Sinemurian | *Caenisites turneri* *Arnioceras semicostatum* | IV | Shales with Beef 21·33m | Lower Lias |
| Hettangian | *Arietites bucklandi* *Schlotheimia angulata* *Alsatites liasicus* *Psiloceras planorbis* | III  II  I | Blue Lias 32m | Lower Lias |

bituminous paper shales with calcareous nodules each invested with a film of beef. At the top the Birchi Nodule Bed is one of the few beds in this unit that yields well-preserved fossils, especially the ammonite *Microderoceras birchi*. The Shales-with-Beef can be studied beneath Black Ven. The Birchi Nodule Bed is at beach level on the right bank of the mouth of the River Char.

The only continuous bed of limestone in the Shales-with-Beef is the Black Arnioceras Bed some 7 m below the top. This general absence of limestone suggests deposition in deeper water than the Blue Lias, and the paper shales with poorly preserved fossils suggest a progressive deepening.

The Black Ven Marls are well seen in the cliffs of Black Ven, and in Stonebarrow Cliff east of Charmouth. They consist of dark shales with impersistent layers of limestone which, where fossiliferous, yield magnificently preserved ammonites, e.g. *Caenisites brooki, Promicroceras planicosta, Asteroceras stellare* and *Echioceras raricostatum*. Bottom-living (benthonic) forms are rare.

Three beds in the Black Ven Marls are of special interest: the Pentacrinite Bed, 25 mm thick or less, and seen as an impersistent layer on Black Ven, yields magnificent specimens of the crinoid *Pentacrinites fossilis*. Some 3.5 m higher, abundant fossil insects have been found—mainly grasshoppers and aquatic beetles. About 5 m above this level is the Coinstone. This is a layer of flat nodules that have been lithified, eroded, encrusted and bored into by organisms, and pyritised all in the space of time represented by the top part of the *obtusum* Zone and the *oxynotum* Zone, which are missing here (Hallam, 1969). Beds formed in this way are termed 'hardgrounds' and indicate non-deposition (perhaps also removal) of sediment. The Coinstone is the best example in the Dorset Lower Lias. The top bed of the Black Ven Marls, a limestone named Hummocky, with abundant ammonites (*Echioceras*) on its under surface, marks a non-sequence in the *raricostatum* Zone.

The Black Ven Marls are thought to have been deposited in shallow lagoons sheltered from waves and currents by offshore bars and by dense growths of seaweed which would have prevented oxygen and daylight from penetrating far below the surface of the sea.

The next higher division, the Belemnite Marls, appears as a well-marked pale grey band in the cliffs of Black Ven and Stonebarrow Hill, and on the foreshore below Golden Cap. The beds consist of alternate layers of darker, bluish grey, and paler, greyish buff marls; the paler bands are more calcareous than the darker ones. The fact that both dark and pale layers are extensively mottled by the trace-fossil *Chondrites* shows that the alternation is primary (Sellwood, 1970). Belemnites are common throughout, but especially in the top bed, a limestone 0.2 m thick called the Belemnite Stone. Ammonites (*Apoderoceras, Platypleuroceras, Tropidoceras, Beaniceras*) are usually found pyritised and are better preserved in the upper than in the lower part.

The Green Ammonite Beds, so named from the colour of the calcite found filling the chambers of some of the ammonites, occur in the cliffs of Stonebarrow Hill; they are thicker and more easily studied between St Gabriel's Water and Seatown. They are pale grey and buff clays, less calcareous than the Belemnite Marls, and become silty and micaceous in their upper part. About halfway up, a composite bed of reddish-weathering limestone 1.7 m thick, the Red Band, separates beds with *Aegoceras (Androgynoceras)* below from beds with *Oistoceras* above.

The Belemnite Marls and Green Ammonite Beds yield a more abundant and diverse benthonic fauna of brachiopods, bivalves and gastropods than do the underlying beds of the Lower Lias, and are more bioturbated. This shows that they were laid down in water which, if not shallower, was certainly better lit and oxygenated.

**Plate 1** Fossils from the Lower and Middle Lias
1 *Microderoceras birchi, a* side view, *b* ventral view. 2 *Echioceras raricostatum*. 3. *Amaltheus margaritatus*. 4 *Plagiostoma giganteum*. 5. *Quadratirhynchia quadrata, a* dorsal view, *b* anterior view. 6 *Aegoceras (Androgynoceras) lataecosta*. 7 *Calcirhynchia calcaria, a* dorsal view, *b* anterior view. 8 *Liostrea irregularis*. 9 *Pseudohastites sp.*

Jurassic 15

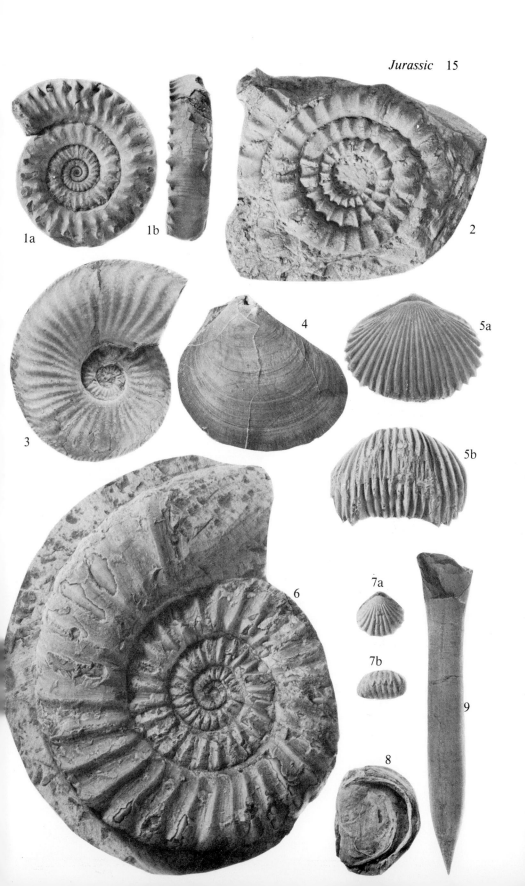

## Middle Lias

The Middle Lias is exposed on the eastern side of Stonebarrow Hill, on Golden Cap and on Thorncombe Beacon. It consists, broadly speaking, of some 122 m of silty micaceous clays becoming more sandy towards the top, with certain beds of distinctive lithology and fauna mentioned below. The beds vary in thickness, some thickening to the east and others to the west. The thicknesses cited here are generally maxima.

The base of the Middle Lias is 3.35 m below three thick bands of calcareous sandstone, the Three Tiers, which form projecting buttresses along the lower parts of Golden Cap. They are succeeded by bluish grey clays, the Eype Clays or Micaceous Beds. At 38 m above the base of the latter is the Eype Nodule Bed, a layer of intensely hard nodules of calcite-mudstone showing evidence of two phases of calcite segregation. Ammonites are crushed in the clays but well preserved in the nodules. About 20.25 m higher is Day's Shell Bed (0.25 m), a sandy marl, indurated in some places but friable elsewhere, and packed with crinoid ossicles. Palmer (1966b) showed that its fauna (60 to 70 species, about half of them bivalves) was swept together by currents that winnowed away an originally greater thickness of sediments from a wide area—for it appears that benthonic animals could not survive the conditions of the Eype Clays.

Next in the succession are the Laminated Beds or Down Cliff Sands (26.21 m), which consist of blue, brown and grey micaceous sands and clays with bands of sandstone and of ironstone nodules. At the base is the Starfish Bed, a hard greenish grey micaceous sandstone on the under side of which are found complete skeletons of the brittle-star *Ophioderma*. The sands contain the brachiopod *Gibbirhynchia muirwoodae* and the bivalves *Gryphaea cymbium* and *Pseudopecten equivalvis*. These beds are capped by the Margaritatus Stone, about 0.3 m of massive grey sandy limestone yielding *Amaltheus subnodosus, A. margaritatus* (Plate 1) and numerous small gastropods. Above it is the Margaritatus Clay (2.1 m), with the same ammonites and 17 species of bivalves not found in the stone. Most specimens are crushed and decalcified.

The Thorncombe Sands, up to 26.68 m of yellow sands, contain doggers at various levels, some showing *Thalassinoides* on their under surfaces. These contain species of *Gibbirhynchia* and *Amaltheus*. The sands are locally capped by a 35-cm bed packed with *Gibbirhynchia thorncombiensis* and followed by some 2.9 m of marly sands.

The top of the Middle Lias is marked by the top of the Marlstone Rock-Bed, a greenish grey, brown-weathering limestone with limonite pellets, only 20 cm thick on the coast (where it forms part of the Junction Bed described below) but thicker and highly fossiliferous inland. Characteristic fossils include the ammonite *Pleuroceras* and the brachiopod *Quadratirhynchia*.

Inland, the Middle Lias crops out north-west of Sherborne; but because none of the distinctive beds seen on the coast is developed, the base is ill defined within an upward transition from the silty top of the Lower Lias. Two divisions can, however, be recognised: the Middle Lias Marls (15.25 m) below and the Pennard Sands (up to 24.3 m) above. These are the attenuated representatives, respectively, of the Eype Clays and the combined Down Cliff and

Thorncombe sands. The Marlstone Rock-bed (0.4 to 4.25 m) is a grey to brown ferruginous limestone with a more varied brachiopod fauna and many belemnites in addition to the fossils found on the coast.

The gradual upward passage from micaceous silty clays to sands and finally to limestone indicates a progressive shallowing of the sea which continued in Upper Lias times.

## Upper Lias

Over most of the area of north-west Europe, the beginning of Toarcian time was marked by a rise of sea level which reached its maximum in *falciferum* Zone times. In the area, however, there was formed one of the most remarkable deposits in the British Jurassic—a limestone rarely more than 1.5 m thick, and corresponding to over 91 m of strata in Yorkshire, in which are represented the lowest four-and-a-half zones of the Toarcian (see Table 3). This bed forms, with the Marlstone Rock-Bed to which it is usually welded, the Middle–Upper Lias Junction Bed, which was studied in great detail by Jackson (1926).

The main part of the Upper Lias portion of the Junction Bed forms three layers, with a fourth, basal layer of *tenuicostatum* Zone age in some places. The limestone is cream coloured, yellow, pink or grey and has a smooth fracture. In places it is mottled with iron or manganese oxides. There are planes of non-sequence within and between the layers, and the abundant ammonites lie at all angles in the rock. In some places the layers are separated by black clay, varying in thickness from a mere film to several centimetres.

These characteristics indicate conditions of deposition found in many Jurassic limestones deposited in the Mediterranean area (Tethys of Jurassic times) and named 'ammonitico rosso' from their occurrences in Italy. Sediment accumulated extremely slowly, trapped by mats of blue-green algae whose characteristic encrustations, or stromatolites, film surfaces within and between the layers and encrust some of the ammonites. Erosion may have caused the non-sequences in the deposit; but it is also possible that acids secreted by the algae corroded much of the sediment and its fossils, leaving an insoluble residue of black clay. Since algae need light, it is obvious that the beds were deposited in quite shallow water. But the amount of land-derived sediment is negligible. The Tethyan 'ammonitico rosso' limestones are held to have been formed on sea-mounts far out in the open ocean; but in the case of the Junction Bed, abundant terrigenous deposits were accumulating at the same time within 160 km to the north. It is thus not yet possible to account for the fact that there was virtually no subsidence in the area of the Junction Bed for several million years. Jenkyns and Senior (1977) interpret fissure-fillings and changes in thickness of the Junction Bed at Eype Mouth as evidence of fault movement during deposition.

The top of the Junction Bed is certainly eroded and marks a non-sequence. On it, in the southern part of the area only, rests the Down Cliff Clay which yields ammonites (*Dumortieria*) of the upper half of the *levesquei* Zone. This clay is 21.34 m thick on the coast but passes laterally into Bridport Sands just to the north.

While the Junction Bed was slowly accumulating in the south, a sandy deposit in the form of an offshore bar was advancing from the north. This began in the north Cotswolds in *bifrons* Zone times and entered the area near Bath in late *thouarsense* Zone times; it is here called the Midford Sands. We next see it as the Yeovil Sands, of *levesquei* Zone age, near Sherborne and Yeovil—and notably at Ham Hill, where a local limestone 24.3 m thick, the Ham Hill Stone, lies within the sands and yields *Dumortieria moorei* (Plate 2). Near Beaminster the sands reach their maximum thickness of 91.5 m. On the coast, as the Bridport Sands, they yield *Pleydellia aalensis*, the ammonite of the last subzone of the *levesquei* Zone and of the Toarcian Stage, with species of *Leioceras* proving the *opalinum* Zone of the Bajocian, which is part of the Inferior Oolite inland.

The Midford, Yeovil and Bridport sands consist of alternate beds of fine-grained sand or crumbly sandstone, and sandy limestone. Their mode of formation has been studied by Davies (1967; 1969). The sand beds vary from 30 cm to 3 m in thickness and become progressively thinner up the sequence; the sandy limestone beds range up to 75 cm and have irregular upper and lower surfaces (Plate 4.1). Shelly fossils are not common, but many of the beds have been thoroughly churned and burrowed by animals living on or within the sea-bed, possibly in the inter-tidal zone.

Davies attributed the progressive thinning of the sandy beds to a steady diminution of the supply of sediment with a constant supply of calcium carbonate. The supply of sediment slowly decreased as the area of sand deposition moved southwards until the sands were succeeded by limestones with only a small detrital content (Cephalopod Bed of the Upper Lias north of the Mendips; Inferior Oolite south of the Mendips).

The Ham Hill Stone was formed in a tidal channel which cut through the sand bar to the back-bar area (Figure 4). Strong currents deposited shell-fragment limestones in a north-easterly direction, although the main sand bar manifestly advanced in the opposite direction.

## Middle Jurassic

The Middle Jurassic was a period of rapid local variation in the type and thickness of sediment and of regression of the sea. The Inferior Oolite represents a time of shallow, often turbulent seas. In Great Oolite times, although a clay formation was the first to be laid down, these calmer conditions again gave way to turbulent seas, first in the north and later over the whole region.

### Inferior Oolite

The beds of the Inferior Oolite vary considerably in lithology, but are all limestones in this region—oolitic (in many cases ferruginous), shell-fragmental oolitic, or sandy. A soft building stone occurs at Sherborne; other beds are used for road metal, stone walls and lime burning.

The term 'oolite' (Greek *oon*, egg) denotes rock that resembles the roe of a fish in texture because it consists mainly of small spheroidal grains. Fossils—mostly ammonites, other molluscs and brachiopods—are abundant. When broken up, their debris constitutes shell sand which may form beds described as sandy even though they contain no silica. There is evidence of earth movements that caused gentle folding of the sea floor. Currents then eroded the crests of the folds and new deposits were laid down over the beds

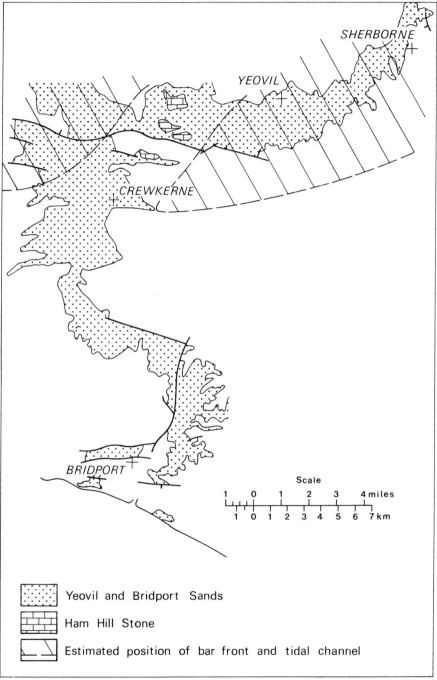

**Figure 4** Outcrop of the Yeovil and Bridport Sands showing the estimated position of the bar front and tidal channel at the time of formation of the Ham Hill Stone (After Davies, 1969)

thus laid bare. Thus not only does the thickness of the Inferior Oolite vary, but beds presumably originally deposited over wide areas are now known from only a few places. Lower, Middle and Upper divisions of the Inferior Oolite are recognised. The planes between them mark the erosion that accompanied minor transgression of the sea.

In Dorset the Inferior Oolite is interesting for the abundance and diversity of its fossils and for the light they throw on the conditions of deposition. S. S. Buckman (1922), by showing how ammonites could be used to unravel the intricacies of these highly condensed deposits, laid the foundations of modern biostratigraphy.

The base of the Inferior Oolite is taken at the base of the bed marking the replacement of Yeovil–Bridport sands deposition by normal limestone. In the Yeovil–Sherborne area this change takes place at the Dew Bed, a hard shelly limestone with much crystalline calcite which yields Toarcian ammonites of the *levesquei* Zone. On the coast, on the other hand, the change occurs within the *scissum* Zone of the Bajocian.

The Inferior Oolite is generally thin at outcrop in this region: about 3.6 m at Burton Bradstock; 6 m around Chideock and Crewkerne; 1.8 m or less at Haselbury and Yeovil Junction; 18 m at Sherborne, and about 7.6 m where next seen in this area north of Box. East of the outcrop, however, it thickens rapidly: 39.3 m at Westbury; 44.5 m at Stalbridge; 72.8 m at Wincanton; and 120 m at Kingsclere. In these thicker sections it is still a limestone formation.

There was a general westward transgression of the sea in early Bajocian times. This was followed by a regression and by gentle folding of the sea floor which was raised locally above wave base, so that the beds were eroded away, in some places down to the Upper Lias sands. Thus the Lower and Middle Inferior Oolite are variable, and nowhere completely developed at outcrop. There followed a general advance of the sea—the 'Bajocian transgression' —over a front from Dorset to the north Cotswolds. In the region the Upper Inferior Oolite commonly forms more than half, and in some places (between Crewkerne and North Coker; and north of Box) the whole of the formation. In Dorset it generally forms thicker beds of more uniform lithology and with fewer non-sequences than the lower divisions.

Some condensed beds in the Lower and Middle Inferior Oolite are thin, ironshot limestones rich in well-preserved ammonites. Detailed subzonal sequences can be worked out in such beds (Parsons, 1976). Others consist largely of conglomeratic accumulations of ammonites, many of them abraded, from several horizons. Such beds, which defy detailed analysis, may also contain disc-shaped limonitic stromatolites 1 to 30 cm in diameter called 'snuffboxes'. These were originally growths of blue-green algae which served as sediment traps at times and places of very slow deposition. Although the algae were not skeleton-secreting types, limonite replacements of their calcified

**Plate 2** Fossils from the Upper Lias, Inferior Oolite, Fuller's Earth and Bradford Clay

1 *Dumortieria moorei.* 2 *Liostrea hebridica.* 3 *Graphoceras decorum.* 4 *Neocrassina modiolaris,* *a* right valve, *b* hinge of same. 5 *Liostrea hebridica* subsp. *elongata.* 6 *Apiocrinus parkinsoni.* 7 *Rhynchonelloidella smithi, a* dorsal view, *b* anterior view. 8 *Goniorhynchia boueti, a* dorsal view, *b* anterior view. 9 *Parkinsonia parkinsoni.* 10 *Digonella digona, a* dorsal view, *b* anterior view. 11 *Wattonithyris wattonensis, a* dorsal view, *b* anterior view.

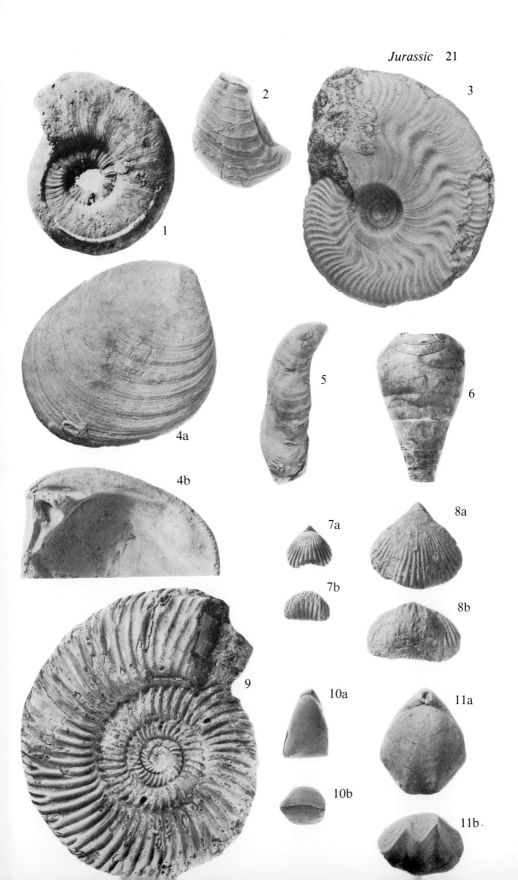

filaments may sometimes be found. The upper surfaces of these beds are usually planed off, and cut through sediment, fossils and 'snuff-boxes' indiscriminately. Burrows made before the sediment hardened may be filled by sediment from the overlying bed, or by material no longer represented in any other way at that locality. The analogy with the Junction Bed at the base of the Upper Lias is obvious.

The best-known section of the Inferior Oolite is at East Cliff, Burton Bradstock (Figure 5). Another well-known section is a road-cutting beside the A30 road at Halfway House, between Yeovil and Sherborne (Figure 6), a section that warrants preservation.

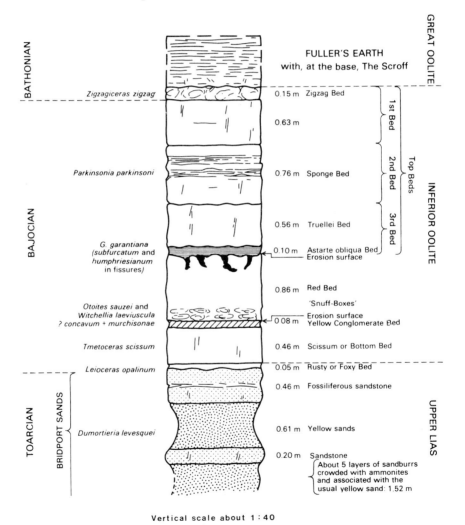

**Figure 5** Vertical section through the Inferior Oolite at Burton Bradstock (Adapted from Richardson, 1928, p.60)

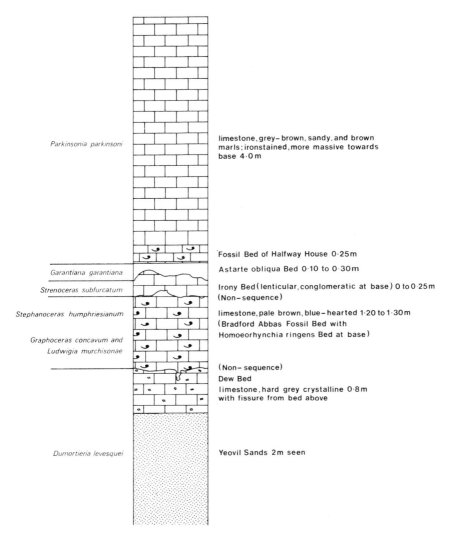

**Figure 6** Vertical section through the Inferior Oolite at Halfway House

## Great Oolite Series

The Great Oolite Series was named after its most distinctive member, the massive oolitic limestones of the Bath area which are the source of Bath Stone. However, these limestones occupy only a minute part of the area of outcrop of the Series in this region. A line running from west to east at about the latitude of Westbury and then turning to run southerly under Poole Harbour separates two areas of different histories. South and west of this line, in the Dorset Basin, the Series consists largely of clays: the Fuller's Earth is succeeded by the Frome Clay and the Forest Marble; the lower formations consist almost entirely, and the upper one largely, of clay (though with beds of limestone). North and east of the line, the Fuller's Earth is separated from the

24  *The Hampshire Basin*

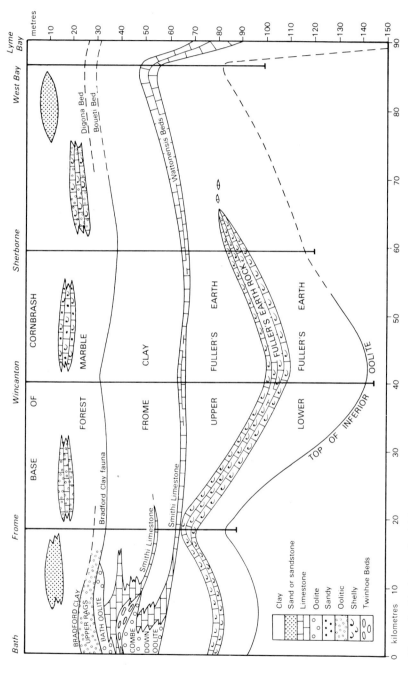

**Figure 7** Variations in thickness and facies in the Great Oolite Series between Bath and Lyme Bay (With acknowledgements to Dr I.E. Penn and Mr R.J. Wyatt)

Forest Marble by the Great Oolite limestones, which replace the intervening Frome Clay. Both sequences are shown in Figure 7, and are described separately below.

Generally speaking this Bathonian Stage (see Table 1) was a period of regression of the sea in southerly and south-easterly directions, and of a reduction in faunal diversity, particularly where the ammonites were concerned. At the very end of the stage, however, there was a renewed marine transgression on at least an epeirogenic scale, and the Cornbrash limestones closed the Middle Jurassic and opened the Upper Jurassic chapter.

The fossils of the Great Oolite Series are facies-controlled. Brachiopods are abundant and, although they do not occur evenly over wide areas, serve to identify certain horizons. The ammonites provide a less comprehensive and reliable framework of correlation than in the Inferior Oolite.

*The Dorset Basin*
At the base of the Lower Fuller's Earth in the Dorset Basin is a hard iron-stained marl called The Scroff, which rests on the Zigzag Bed at the top of the Inferior Oolite. In places The Scroff is replaced northwards by a bank of small oysters with fine ribbing on the left valve. The oysters are *Catinula knorri* and the bed is hence called the Knorri Clays. The overlying Lower Fuller's Earth clay is bluish grey, marly and laminated, with small nodules of pyrite in the lower part; above are grey and brown clays. Fossils are scarce, and of the brachiopods that are common higher in the succession, only rare *Wattonithyris* and *Rhynchonelloidella* are recorded. The most characteristic fossil is *Liostrea acuminata,* which forms a distinct shell bed just below the Fuller's Earth Rock to the north.

In Dorset the Fuller's Earth Rock forms a distinct surface feature where, with a thickness of 10.7 m, it enters the county near Haydon. After some displacement by faulting it reappears in Sherborne Park and extends to the south-west as far as Beaminster, where it has thinned out to a discontinuous line of nodules. In the neighbourhood of Sherborne the rock is a grey or buff coloured argillaceous limestone, weathering to a cream colour. It yields numerous ornithellids and other brachiopods, ammonites (*Tulites*) and bivalves. There is no trace of it between Beaminster and the Dorset coast.

The Wattonensis Beds, rubbly limestones with brachiopods seen on the coast, were formerly thought to represent the Fuller's Earth Rock, but now define the top of the Upper Fuller's Earth. The commonest brachiopods are *Wattonithyris wattonensis* (Plate 2) and species of *Rhynchonelloidella*. Other forms include *Rugitela bullata* and *Tubithyris powerstockensis*, but the ornithellids which are common in the Fuller's Earth Rock are not found. Bivalves, gastropods and proceritid ammonites occur. At Rodden Hive Point, near Langton Herring, these beds yield beautifully preserved bivalves (*Nucula, Myophorella*) and broken ammonites (*Procerites*). Many of the shells are encrusted with worm tubes and polyzoa, showing that the deposit is condensed: foraminifera, ostracods and otoliths may be recovered in profusion. The Wattonensis Beds, which are 7.6 m thick at Watton Cliff, continue inland until they thin away in Sherborne Park, so that between that point and Beaminster there is one rock-band within the Fuller's Earth and another at its top.

The Upper Fuller's Earth clay comprises only the beds between the Fuller's Earth Rock and the Wattonensis Beds. The clays above the latter are now called the Frome Clay (Penn and Wyatt, 1979). This contains a bed full of small gastropods near Cliff End, Burton Bradstock, and a bed crowded with oysters (*Liostrea hebridica elongata*) near Langton Herring.

The combined thickness of the Fuller's Earth and Frome clays at outcrop in the Dorset Basin is not certainly known, because no continuous section is available and the main exposures are affected by faulting. At Watton Cliff it is variously estimated at 91.4 m (Buckman, 1922) and 44.2 m (Wilson and others, 1958). Inland it is about 45.7 m at Powerstock. East of the outcrop, boreholes show that it thickens to over 183 m inland from Langton Herring, nearly 275 m at Kimmeridge, and over 152 m at Bere Regis. The north-eastern limits of the basin are fairly clearly defined, for at Fordingbridge Great Oolite Limestone is present in force and there are but 2.1 m of Lower Fuller's Earth.

The Forest Marble varies from about 29 m between Weymouth and Langton Herring to about 42.7 m around Sherborne. It consists generally of a limestone lying between two clay members, though near the boundary with the Great Oolite province in the north it includes a lenticular sand bed, the Hinton Sands.

In the lower part of the Forest Marble there are two brachiopod beds. The first, the Boueti Bed, defines the base of the formation. On the shore of The Fleet at Herbury it is exposed as a bed of whitish marl packed with fossils, of which many are encrusted with worm tubes and polyzoa, indicating a condensed deposit. Brachiopods are by far the commonest fossils, especially *Goniorhynchia, Avonothyris* and *Digonella,* but bivalves and gastropods also occur and there is a rich microfauna. Ammonites—*Clydoniceras (Delecticeras)*—are very rare. This bed can be traced inland to near Sherborne. The second, the Digona Bed, occurs about 18.3 m higher. *Digonella* replaces *Goniorhynchia*, but otherwise the abundant brachiopods belong to species of the same genera as are found in the Boueti Bed.

The Forest Marble clays are generally greenish or yellowish in colour. They contain thin sandstone 'tiles', the upper surfaces of which are covered with trace-fossils. The commonest of these, *Gyrochorte,* resembles plaited string and is thought to represent the burrows of amphipod crustaceans (Hallam, 1970). The limestones are generally lenticular. They are usually thinly bedded, cross-bedded, and made up of the debris of oysters and other shells. They may contain clay galls and pebbles of the same limestone as the matrix. The rock is very tough and will take a polish (hence Forest 'Marble'). Some beds are fissile and were formerly used for roofing slates. The limestones were deposited in shallow, agitated water. The limited diversity of their fauna suggests sea water of less than normal salinity. The clays indicate sheltered conditions interrupted by occasional heavy storms bringing sand from the nearby land.

The Cornbrash (up to about 9.1 m thick in this region) extends with little interruption from Dorset to Yorkshire. In spite of its lateral persistence, the outcrop is nowhere very wide. The name 'Cornbrash' denotes the stony or brashy nature of the soil to which the rock gives rise at outcrop. It was deposited during two phases of marine transgression. Lower and Upper divisions are recognised by differences in lithology and fossils. In places a pebble bed marks the line of junction and the plane of the second transgression.

The Lower Cornbrash consists chiefly of marly rubble and compact limestone. The Upper Cornbrash includes sandy marl, concretionary limestone and flaggy beds of hard limestone. Among the ammonites the smooth, laterally compressed *Clydoniceras* characterises the lower division, and the robust plain-ribbed *Macrocephalites* the upper. The brachiopods, however, allow two lower and two upper zones to be identified. *Cererithyris intermedia* and *Obovothyris obovata* (Plate 3) characterise the Lower Cornbrash (together with *Kallirhynchia yaxleyensis*), and *Microthyridina siddingtonensis* and *M. lagenalis* (with *Rhynchonelloidella cerealis*) the upper division. Among the bivalves, the small, finely ribbed *Meleagrinella echinata* and species of *Astarte* and *Trigonia* characterise the Lower Cornbrash, and the sharply folded oyster *Lopha marshii* the Upper. The echinoid *Nucleolites clunicularis* is also common.

Douglas and Arkell (1928) described the stratigraphy of the Cornbrash in detail. It is now poorly exposed, though both divisions can be seen at Chesters Hill on the shore of The Fleet near Abbotsbury Swannery.

*Northwards from Westbury*
The Great Oolite Series passes out of the area between Sherborne and Milborne Port. Where we see it again, near Bradford-on-Avon, it has undergone great changes, for we are here in the narrow belt of transition from the mainly clay facies of Dorset to the mainly limestone facies of the Cotswolds. The complexities of this transition have been expounded by Penn and Wyatt (1979).

The Lower Fuller's Earth, which is 36.5 m thick in a borehole at Wincanton, has thinned to 14 m at Frome. From here to Bath (and beyond) it varies little in thickness and consists of a number of consistent beds whose lithology is much the same as in south Dorset. Penn and Wyatt showed that it was deposited in three cycles, each of which started with a relatively rapid shallowing of the sea (represented by a shelly bed) following a pause in sedimentation, the inverse of the course of events in Lower Lias cycles. Each pause was followed by a slow deepening of the sea until the start of the next cycle. They attributed this to the behaviour of a stable or 'positive' structural belt corresponding roughly with the eastward continuation of the line of the Mendips (the 'Mendip Axis' of earlier authors). Conditions were those of a low-energy shelf sea directly connected to the open ocean.

The Fuller's Earth Rock, 3.4 to 5 m thick, is a rubbly, shelly, argillaceous limestone and rests with an erosive base on the Lower Fuller's Earth. It contains two rich brachiopod faunas in the upper part. The lower one is the more widespread and is dominated by *Ornithella*; the higher is characterised by *Rugitela*, which, however, is only locally dominant.

It is in the beds between the Fuller's Earth Rock and the Forest Marble that the most striking changes take place. The Upper Fuller's Earth is 28 m thick at Bath. Here the bed of commercial fuller's earth from which the formation is named is 3.25 m thick and lies about 10 m below the top; it has thinned to nothing within 10 km to the south. Jeans and others (1977) have shown that the earth is the product of re-working of volcanic ash first deposited nearly contemporaneously elsewhere in the vicinity. At Frome the whole of the Upper Fuller's Earth is reduced to 0.5 m and is separated from the succeeding

Frome Clay by a thin bed of argillaceous limestone with *Rhynchonelloidella smithi* (Plate 2). Penn and Wyatt showed that there is a hiatus between the Upper Fuller's Earth and this Smithi Limestone. The Frome Clay (about 30 m thick at Frome) is a mudstone like the Upper Fuller's Earth. In the type area it can be divided into three parts by reference to a second Smithi Limestone about 10 m above the first, and a bed some 15 m higher again in which the clay is extensively burrowed.

In a belt about 3.2 km wide at the latitude of Faulkland and Baggridge Hill the Frome Clay passes into the limestones of the Great Oolite (Plate 5.1). These consist, in upward succession, of the Combe Down Oolite, the Twinhoe Beds and the Bath Oolite (the first and third are the source of the Bath Stone). The basal Smithi Limestone of the Frome Clay passes into the uneven, erosive base of the Combe Down Oolite. The second Smithi Limestone passes into the ironshot limestones at the base of the Twinhoe Beds. The burrowed bed passes into shelly mudstones and limestones which can be recognised in boreholes at the base of the Bath Oolite around Hinton Charterhouse. The many facies variations in the limestones were discussed by Penn and Wyatt; the general stratigraphical relationships are shown in Figure 7.

The oolites were deposited in two very shallow-water environments. There are, first, flat-bedded limestones in which each bed—from 0.3 to 1.5 m thick—is normally coarser grained below and finer grained above. These were deposited rapidly by waters advancing across flats covered only at high tide. Cross-bedded layers represent the migration of sand waves or the accumulation of shell banks. The top of each bed in most cases forms a hardground plastered with oysters and bored by bottom-living organisms after lithification, and clearly marks a pause in deposition. Secondly, there are limestones, usually packed with shells, in which the bedding is of the type termed 'trough cross-bedding'. These were formed quickly in tidal channels. Non-oolitic, silty or marly limestones are found locally; they were deposited in lagoons sheltered by shell-banks.

The Twinhoe Beds, which display various facies, were formed on a gently sloping floor where the sea deepened from the tidal flats and channels of the oolites to depths where the mudstones of the Frome Clay continued to accumulate.

The base of the Forest Marble in this area is represented in places by a limestone (the Upper Rags), and elsewhere by a shell bed containing the Bradford Clay fauna. This fauna, which is highly distinctive, occurs at more than one horizon, but the two most prominent beds are correlated with the Digona Bed and Boueti Bed of the Dorset coast. These shell beds include a wealth of brachiopods (*Digonella, Avonothyris, Epithyris* and *Rhactorhynchia* are common; the more striking *Eudesia* and *Dictyothyris* are scarce), bivalves (*Chlamys, Oxytoma*), and the pear-shaped crinoid *Apiocrinus*. The basal junction is always sharply defined. Where the underlying bed is a limestone, its top is commonly an uneven hardground with hollows eroded back to pro-

**Plate 3** Fossils from the Cornbrash and Oxford Clay
1 *Quenstedtoceras lamberti, a* side view, *b* ventral view. 2 *Cardioceras* cf. *cardia, a* side view, *b* ventral view. 3 *Cererithyris intermedia, a* dorsal view, *b* anterior view. 4 *Meleagrinella echinata*. 5 *Microthyridina lagenalis*. 6 *Gryphaea dilatata*. 7 *Obovothyris obovata, a* dorsal view, *b* anterior view.

Jurassic 29

duce overhangs. The hardground and both the ceilings and floors of the hollows are bored and encrusted with worm-tubes, polyzoa, attached bivalves (oysters and *Plicatula*) and *Apiocrinus* (Palmer and Fuersich, 1974). The Bradford Clay and Upper Rags that cover the hardground to a thickness of up to 3 m contain the same brachiopod fauna and detached ossicles of *Apiocrinus*. The Upper Rags around Bath include the Corsham Coral Bed at the base and the Bradford Coral Bed at the top, but these are little more than local patch reefs. Between and around them is the Ancliff Oolite, of a similar facies to the Bath Oolite, but containing elements—mainly minute bivalves and gastropods—of the Bradford Clay fauna.

The main mass of the Forest Marble is highly variable in this area. Within a mainly clay formation, sandy and calcareous components range from mere wisps to lenses 9 m thick of sand with doggers or more continuous beds of flaggy or of shelly oolitic limestone.

In the Cornbrash, part of the lower division is a hard blue-centred limestone with few fossils (Corston Beds); the upper division diminishes northwards to a thin ferruginous marl.

## Upper Jurassic

The Upper Jurassic sea, already enlarged by the Callovian transgression represented by the Upper Cornbrash, continued to deepen through Oxford Clay times. There then followed an episode of cyclic regressions and transgressions represented in the changing facies of the Corallian Beds. The Jurassic basin of deposition reached its widest extent and greatest depth during the deposition of the Kimmeridge Clay, by which time the tectonic movements referred to on p. 8 had practically ceased. The cyclic regressions and transgressions of the Portland Group mark the gradual filling up of the basin and the virtual exclusion of the sea in Purbeck Group (Lulworth Formation) times.

**Kellaways Beds and Oxford Clay**
The Cornbrash passes gradually up into the Kellaways Clay, green, grey or blue clays with nodules and silty layers, up to 20 m thick. Above the clay is a sandy limestone, the Kellaways Rock, irregularly developed in extent and varying in thickness from 0.5 to 4 m. The ammonite fauna of these two divisions of the Kellaways Beds (Table 4) includes the genera *Cadoceras, Chamoussetia, Kepplerites, Macrocephalites* and *Sigaloceras*. The numerous bivalves include *Anisocardia, Corbulomima,* oysters and *Oxytoma*. Belemnites and brachiopods are common in the Kellaways Rock. The beds are scarcely exposed in the region except intermittently on The Fleet shore south of East Fleet church, but there are good sections in the banks of the Avon just beyond the northern boundary and south of Kellaways village.

The Oxford Clay above the Kellaways Beds shows a nearly uniform sequence of clay and shale lithologies over its whole extent. Its outcrop is continuous (except for the Chalk overstep in north Dorset) from the Dorset coast to Yorkshire. Its thickness varies in the area from 120 to 155 m on the coast to 183 m in north Wiltshire. In Dorset it is poorly exposed along the shore of the East Fleet, where all the zones of the Oxford Clay are exposed from time to

**Table 4** Stratigraphy of the Kellaways Beds and Oxford Clay

| STAGE | AMMONITE ZONE | LITHOSTRATIGRAPHICAL DIVISION | |
|---|---|---|---|
| OXFORDIAN (PARS) | *Cardioceras cordatum* | CORALLIAN BEDS (PARS) | |
| | | RED NODULE BEDS | |
| | | JORDAN CLIFF CLAYS | |
| | *Quenstedtoceras mariae* | FURZEDOWN CLAY | OXFORD CLAY (Estimated 150m) |
| CALLOVIAN | *Quenstedtoceras lamberti* | (Clays and shales, bituminous in part with septarian concretions) | |
| | *Peltoceras athleta* | | |
| | *Erymnoceras coronatum* | | |
| | *Kosmoceras jason* | | |
| | *Sigaloceras calloviense* | KELLAWAYS ROCK (0·5 to 4·5 m) | KELLAWAYS BEDS (3 to 25m) |
| | | KELLAWAYS CLAY (2·5 to 21m) | |
| | *Macrocephalites macrocephalus* | UPPER CORNBRASH | |

time on either side of Tidmoor Point. The most fossiliferous section is at the point itself, where abundant pyritised ammonites of the *lamberti* Zone (Arkell, 1947, p. 31) are washed out by wave action. The same beds are exposed farther east at Ham Cliff. The younger Furzedown Clays of the *mariae* Zone, again with pyritised ammonites, occur at the point of that name on the Fleet shore (Arkell, 1947, p. 32) and also north-east of Weymouth between Redcliff Point and Shortlake. The succeeding Jordan Cliff Clays and Red Nodule Beds of the *cordatum* Zone are best exposed in Jordan or Furzy Cliff, east of Weymouth; the rusty-coloured flat nodules of the latter, commonly containing ammonites (*Cardioceras*), are conspicuous on the beach. The top 12 m or so of the Oxford Clay are poorly exposed. There are no permanent sections at this level inland, but the lower beds of the Oxford Clay are exposed at Crook Hill Brickyard, Chickerell, where brown and blue clays overlie brown and green bituminous shales with crushed ammonites (mainly *Kosmoceras*) and septarian nodules.

**Corallian Beds**

The full sequence of the Corallian Beds, reaching a thickness of about 61 m, is well exposed on the coast between Weymouth and Ringstead (Figures 8 and 9); the succession formerly seen in the cliffs south of Weymouth has been largely obscured by building and by vegetation as a result of reduced coastal erosion behind Portland Harbour. The name of the formation was adopted from French vernacular usage for coral-rich limestones of approximately the same age, but is not descriptive of the beds seen on the coast.

The ammonite zones in Figure 9 are those traditionally used in Dorset. An alternative scheme has been put forward by Sykes and Surlyk (1976), but its application in this region has not yet been worked out.

As pointed out by Arkell, these beds represent three successive cycles of deposition in progressively shallowing water (Talbot, 1973, recognised four cycles), each beginning with clay, continuing with sand and ending with limestones, oolitic in parts. On the coast, corals occur only in the Ringstead Coral Bed, although certain beds in the Osmington Oolite consist of coral debris reduced to silt grade. Patch reefs occur at Steeple Ashton, near Westbury.

| *First Cycle* | *Second Cycle* | *Third Cycle* |
|---|---|---|
| 3 Trigonia hudlestoni Bed (corals in Wiltshire) | 3 Osmington Oolite and T. clavellata Beds (coral patch-reefs in Wiltshire) | 3 Westbury Iron Ore and Ringstead Coral Bed |
| 2 Nothe Grits | 2 Bencliff Grit | 2 Sandsfoot Grit |
| 1 Oxford Clay | 1 Nothe Clay | 1 Sandsfoot Clay |

The Nothe Grits, 3.6 to 10.6 m thick, are grey, thoroughly bioturbated sandstones with 'cannon-ball' concretions in some beds. Fossils include *Gryphaea dilatata* and *Cardioceras*. The Trigonia hudlestoni Bed, 1.82 m thick, is a brownish grey, massive or flaggy, sandy bioturbated limestone and is very fossiliferous (*Myophorella* [*Trigonia*] *hudlestoni*, many other bivalves, *Perisphinctes, Cardioceras* and gastropods).

1 Cliffs of Upper Lias, Bridport Sands, West Bay, showing bands of calcareous sandstone weathered out by wind-blown sand. Note progressive upward thinning of soft beds. (A.12038)

2 Cliffs of Lower Lias, Blue Lias, east of Lyme Regis, showing alternation of limestone and shale. (A.12036)

Plate 4

**1** Quarry face in Great Oolite limestones showing Twinhoe Beds (well-bedded limestone) resting on Combe Down Oolite, Winsley Quarry, near Bath. (A.9734)

**2.** Cliffs west of Kimmeridge Bay showing Kimmeridge Clay with, in foreground, the hard ledge of Broad Bench. (A.12230)

Plate 5

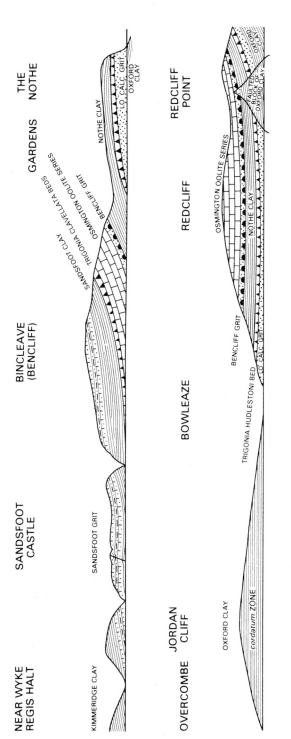

**Figure 8** Section along the cliffs from south of Weymouth north-east to Redcliff Point (From Arkell, 1933, p. 382)

## 36 The Hampshire Basin

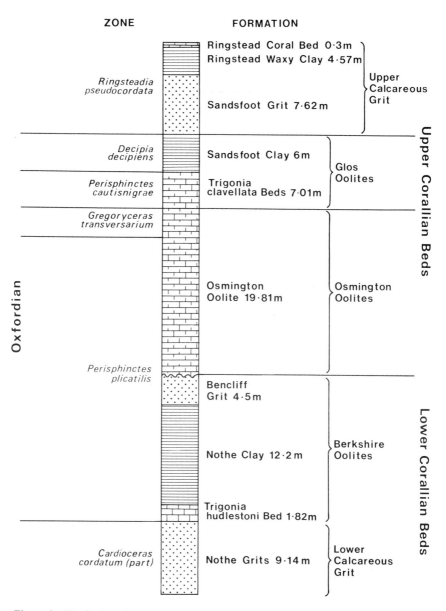

**Figure 9** Vertical section through the Corallian Beds east of Osmington Mills (Adapted from Torrens (editor), 1969, fig. A24)

The Nothe Clay, up to 12.2 m thick, is bluish grey and sandy. It contains limestone bands which are nodular in places. Its fossils include *Gryphaea dilatata* and burrowing bivalves such as *Pleuromya*. The Bencliff Grit (3 to 4.5 m) consists of yellowish, trough cross-bedded sands and sandstones with large doggers. Some beds are strongly impregnated with oil; seepages may be seen on the foreshore east of Osmington Mills. At the junction with the overlying Osmington Oolite there are signs of downcutting and non-sequence with marl-filled burrows in the top few centimetres of the sands, and pebbles of the sands in the basal bed of the oolite. This marks the junction of the Berkshire Oolites and Osmington Oolites of Arkell (1947), and correlates with that between the Lower and Upper Corallian Beds of H. B. Woodward.

The Osmington Oolite (up to about 20 m) comprises a mixture of lithologies—white oolite, grey oolitic marl, nodular limestone and pisolite. Some beds are crowded with bivalves including *Nanogyra nana* and *Chlamys*. Ammonites, gastropods and echinoids also occur. Trace fossils are well preserved, including *Rhizocorallium* and *Thalassinoides* indicating low-energy environments, and *Diplocraterion* indicating high-energy environments (R. C. L. Wilson, 1968a) (see pp.7 – 8). The Trigonia clavellata Beds (up to about 7 m) include limestones and marls, some beds of which are crowded with well-preserved *Myophorella clavellata* and other bivalves (Figure 10). Ammonites (*Perisphinctes* and rarer *Amoeboceras*) may also be found.

**Figure 10** Trigonia clavellata Beds, Osmington Mills, east of Weymouth, packed with *Myophorella (Vaugonia) clavellata*

The Sandsfoot Clay, 6 to 12.2 m thick, contains the flat, triangular oyster *Deltoideum delta* (Plate 6) in abundance. Above is the Sandsfoot Grit, a sharp, medium to coarse grained, reddish and greenish ferruginous sandstone with many fossils, among them the bivalve *Chlamys midas* and occasionally the ammonite *Ringsteadia*. At the type-locality (Sandsfoot Castle) it is 7.6 m thick, but farther east it may be inseparable from the Sandsfoot Clay. Above this sandstone is another clay bed with thin seams of clay ironstone, nodular in places. This is the Ringstead Waxy Clay, in which *Deltoideum delta* is again common; *Ringsteadia* and serpulids also occur. The upper limit of the Corallian Beds is here marked by the Ringstead Coral Bed, an impersistent greenish marly limestone up to 0.3 m thick, but rich in shells and corals. The corals include *Protoseris waltoni, Thecosmilia annularis* and *Thamnasteria concinna* and among the bivalves are *Ctenostreon proboscideum* and *Camptonectes lens*. Gastropods and echinoid radioles also occur.

The change in conditions of deposition from the sandy and calcareous beds of the Corallian Beds to the clays of the Kimmeridge was gradual in at least the western part of the region, and deposition was slow. The beds of the *baylei* Zone at Ringstead (with *Torquirhynchia inconstans, Nanogyra nana* and *Deltoideum delta*) are condensed. At Abbotsbury the zone is not proved, but the Sandsfoot Grit is followed (perhaps non-sequentially, though not obviously so) by a bed of reddish brown ferruginous sandstone up to 8.5 m thick, the Abbotsbury Ironstone. The high percentage of silica prevented the ore from being worked commercially for any length of time, although a railway was built to transport it. It contains an assemblage of fossils unknown elsewhere in England, including the ammonites *Rasenia* and *Prorasenia* indicating the *cymodoce* Zone, brachiopods (*Ornithella, Aulacothyris*) and a variety of bivalves and gastropods. Brookfield (1973) interpreted the deposit as a south-east facing offshore barrier bar (there is no sign of the proximity of a shoreline), similar to the environment of the sands of the Upper Lias (p. 18). In 1978 he included in the Corallian Beds a coherent unit of mainly sandy sediments, the Passage Beds, which included the Abbotsbury Ironstone and up to 10 m of beds formerly placed in the Kimmeridge Clay.

In a general way these Dorset coast divisions can be recognised inland, but there are changes in lithology. The narrow outcrop can be traced between the vales of the Oxford and Kimmeridge clays from Mappowder, north of the Chalk downs, to the northern limit of the Vale of Wardour, where it is faulted against the Cretaceous beds that bound the Vale. North of Sturminster Newton the Osmington Oolite yields building stones (the Marnhull and Todber freestones). J. K. Wright (1981) has given a full account of the Corallian Beds of north Dorset, with implications for what has been said in this chapter on the interpretation of the Dorset coast exposures.

In the Vale of Pewsey the Corallian Beds extend from near Westbury northwards to Calne. At Westbury, near their top, there is a red and green oolitic rock, the Westbury Ironstone, 3.3 to 4.2 m thick containing *Ringsteadia* with *Deltoideum delta* and other bivalves. At Steeple Ashton a coral reef in the limestone of the second cycle has yielded many well-preserved corals, among them *Thecosmilia annularis* and *Comoseris irradians*. At Calne the Osmington Oolite provides another building stone, the Calne Freestone. Sea-urchins are common fossils here and include *Hemicidaris intermedia* and *Paracidaris florigemma*.

Fuersich (1976, 1977) analysed in detail the relationships between benthonic animals and their substrates in the Corallian Beds. The following broad conclusions may be drawn concerning the area. Clays and silts (e.g.

Nothe Clay, Sandsfoot Clay) were deposited in open shallow seas and yield a highly diverse benthonic fauna with a range of deposit-feeding types. Quartzose silts and sandy deposits (e.g. Nothe Grit) supported a fauna mainly of suspension-feeders, progressively less diverse as the energy of the environment increased (though subsequent dissolution by percolating water may have removed some elements of the original fauna). Such beds were formed on tidal flats and are highly bioturbated except where trough cross-bedded in tidal channels. Coral patch reefs (as at Steeple Ashton) supported an opportunistic fauna of echinoids and boring bivalves; for reasons yet unknown, this fauna is much less diverse than its present-day analogues. Mud-grade and silt-grade limestones (as in parts of the Osmington Oolite) were formed by the attrition of reefs and supported a benthonic fauna less diverse than that of the silty clays and with fewer deposit-feeders. Pure oolites and oolitic iron ores, representing the highest-energy environments, are generally poorly fossiliferous. Condensed marly or sandy limestones, including ferruginous limestones, were deposited either as lime muds in inter-tidal shallow water, where they were stabilised by profuse growths of the sponge *Rhaxella*, or as extensively bioturbated lagoon deposits. Byssally attached, suspension-feeding bivalves such as *Chlamys* show a preference for the hard substrates of these condensed deposits.

The distribution and sequence of the sediments were topographically controlled, according to R. C. L. Wilson (1968a, 1968b). The dominant feature was the Portsdown Swell (named from Portsdown Hill, near Portsmouth), the rising (or non-subsidence) of which inhibited deposition or even led to erosion of earlier-formed beds, first in its south-east part, then later in the north-west. The main shoreline, however, probably ran north-eastwards and lay not far north of the present outcrop in Wiltshire and Berkshire. Sediment was supplied mainly from the north-west into a shallow sea in which strong coastwise currents flowed north-eastwards. The clay and silt formations were probably deposited during a widespread, though slight, deepening of the sea. But the deposition of the sands and limestones which succeed them in each cycle was probably controlled by the rate of erosion of the source-area of the sediments rather than by changes in sea level. The basin was more restricted than that in which the Oxford and Kimmeridge clays were laid down, and deposition apparently kept pace with or outpaced subsidence.

The Corallian Beds thicken eastwards to over 91.4 m, but elsewhere in the area where they are concealed—even in the Isle of Wight—they seldom exceed 45.7 m.

**Kimmeridge Clay**
At the end of Oxfordian times gradual subsidence of the area led to changes in geographical conditions. A long-continued period of clay deposition ensued and resulted in the accumulation of the Kimmeridge Clay. The full thickness of this formation at the type locality was determined only in 1937, when an oil exploration well at Kimmeridge proved a thickness of 177 m below the lowest beds seen in the cliffs, making a total of 503 m. Westwards around Ringstead the formation thins to not much more than 245 m. Farther west it thickens again to about 487 m near Portesham.

Lithological variations are to be seen in the Kimmeridge Clay. In addition to thick clays and shales there are thin bands of oil shale, lime mudstone and silty mudstone, and several prominent beds of limestone and dolomite ('stone bands') of which the highest ones consist almost entirely of coccoliths. Some of these stone bands form dangerous offshore ledges (Plate 5.2 and Figure 11). The stratigraphy of the Kimmeridge Clay is shown in Table 5, which was prepared before the revision by Cox and Gallois (1981). The ammonite zones in the lower part have been studied by Ziegler (1962) and those in the upper part (where final agreement has yet to be reached) by Casey (1967) and Cope (1967, 1978).

The Kimmeridge Clay is well exposed along the coast from Gad Cliff to St Alban's Head; it comes to the surface again, through faulting, to the west at Dungy Head (where it is now obscured by landslips) and at Ringstead and near Osmington Mills, whence it extends inland westwards to Abbotsbury as a narrow tract of flat land between the rising edges of the Corallian Beds to the south and the Portland Group to the north. North of Abbotsbury the Jurassic strata are overstepped by the Upper Greensand with the Chalk above. The Kimmeridge Clay reappears north of the Dorset Downs where the Chalk outcrop swings northwards. Here it forms a fairly broad tract east of the ridge of Corallian Beds until it enters the Vale of Wardour, where its further extension is hidden by overstepping Cretaceous strata and by faulting. Northwards it is covered for 18 km but it crops out again in a broad expanse at the western end of the Vale of Pewsey.

In the main cliff section the most abundant fossils of the clays are ammonites (most commonly *Aulacostephanus* and *Pectinatites*) and bivalves (such as *'Lucina'* and *Protocardia*). Brachiopods, echinoderms, serpulids, gastropods, fishes and plant fragments are locally common.

One of the most interesting lithologies in the Kimmeridge Clay is the kerogen-rich shale (oil shale) found at several horizons, which readily burns, producing a most unpleasant smell. The best known such bed, the Blackstone, occurs in the *wheatleyensis* Zone; it is less than a metre thick and has long been known as the 'Kimmeridge Coal'. It can be traced in the cliff sections near Kimmeridge and also at Ringstead Bay. At the latter locality spontaneous ignition of the shale in 1826 gave rise to a fire that lasted four years, from which the place is known as the Burning Cliff. The outcrop of the shale has also been traced in the Portesham area.

These shales have been tested from time to time with a view to their exploitation as a source of oil, but the high sulphur content is one among several factors that have prevented their being used. Circular pieces of the shale found in the soil near Kimmeridge have long been known as 'coal money'. They are probably the waste of larger pieces that had been turned on a lathe into ornamental objects. The horizon of the Blackstone itself is characterised by small, isolated plates of the pelagic crinoid *Saccocoma*, always preserved as pyrite (Plate 6).

Another noteworthy horizon is the Rotunda Nodule Bed, seen in Chapman's Pool, below Houns-tout Cliff, just west of the stream at beach level. This is a bed of pale, calcareous nodular clay directly beneath a prominent 0.5 m bed of bituminous paper shale. The nodules contain uncrushed, slightly phosphatised ammonites (*Pavlovia rotunda*, Plate 6) up to 0.3 m in diameter, though these are now rare because of over-collecting. Their occurrence here

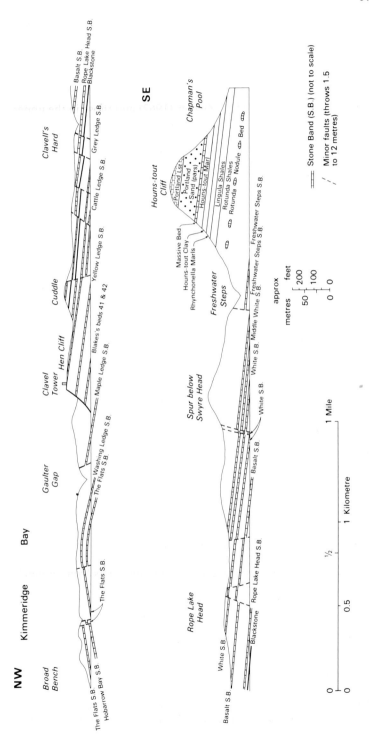

**Figure 11** Section along the cliffs near Kimmeridge Bay, Dorset Walking distance about 8 km. (Adapted from Arkell, 1933, p. 444)

## 42 The Hampshire Basin

**Table 5** Biostratigraphy and lithostratigraphy of the Kimmeridge Clay of Dorset

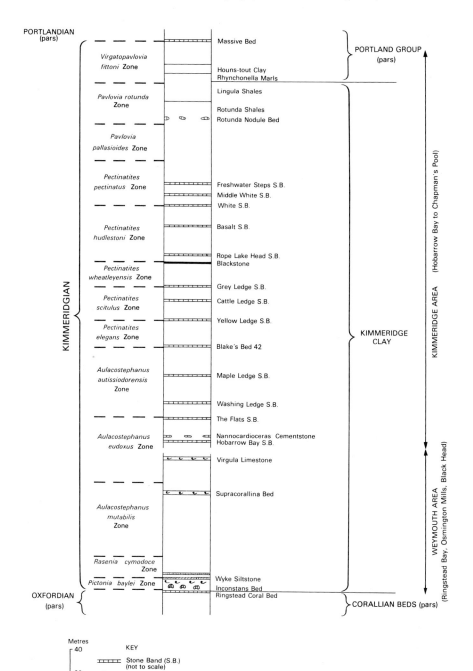

marks only a slight pause in deposition, but at Holworth House, about 18 km to the west, where the Kimmeridge Clay is reduced to half thickness, the ammonites are preserved as small, black, heavily phosphatised nuclei as in the equivalent Upper Lydite Bed of Swindon and Aylesbury which marks a major non-sequence that can be traced at a similar level in Lincolnshire. Casey (1971) described this bed as marking one of the most important events in the British Jurassic from the point of view of long-range correlations and palaeogeography. In this region it marks the first sign of the shrinking of the basin of deposition which culminated in the exclusion of the sea in Purbeck and Wealden times.

Sixteen metres above the Rotunda Nodule Bed is the base of the Lingula Shales, 15 m of dark, well-bedded clays, which mark the end of continuous deposition of clay. They have thus been taken by Townson (1975) to mark the top of the Kimmeridge Clay, and are overlain by the basal beds of the Portland Group, although this boundary is disputed. Cope (1978) has revised the ammonite zones of this part of the succession.

**Portland Group**

The Portland Group was deposited during a period of regression of the sea (probably of epeirogenic dimensions, following Hallam and Sellwood, 1976), interrupted by at least two minor transgressions. Townson's (1975) revision of the lithostratigraphic classification is followed here with the revised ammonite zones of Wimbledon and Cope (1978). The beds were deposited in two parts of a single basin, respectively east and west of a swell which trended north-east–south-west and plunged southwards towards the Isle of Portland from a line through Ringstead Bay and Lulworth Cove. The Group comprises the Portland Sand Formation below and the Portland Limestone Formation above. Townson's classification is compared with that formerly used (Arkell, 1947) in Figure 12 and his correlation is shown in Figure 13.

The base of the Portland Group at Houns-tout Cliff (where the succession is thickest) is now taken at the base of the Rhynchonella Marls (and their presumed correlatives elsewhere), a maximum of 34 m below Arkell's base at the Massive Bed there. The Rhynchonella Marls differ from the shaly clays of the Lingula Shales beneath in being silty mudstones, becoming more silty upwards. They form the lowest part of the Black Nore Member, which consists of these mudstones interbedded with calcareous siltstones, fine sandstones and calcareous silty mudstones. The lithology of the Black Nore Member, though highly variable, is recognisable over the whole area, but in the 'West Basin' the member is more clearly defined by being siltier and glauconitic. The silty and sandy component is, however, not quartzose, but dolomitic, consisting of the ferruginous dolomite called ankerite. The implications of this are considered below.

The Massive Bed is a local calcareous siltstone in the upper part of the Black Nore Member, and it can hardly be identified at Gad Cliff, only 8 km to the west. Moreover, even at Houns-tout, the beds above it do not differ significantly in lithology from those below. Townson's base of the Portland Group corresponds closely with that mapped on the six-inch scale by the Geological Survey.

Fossils are rare and poorly preserved in the Black Nore Member. They

# 44  The Hampshire Basin

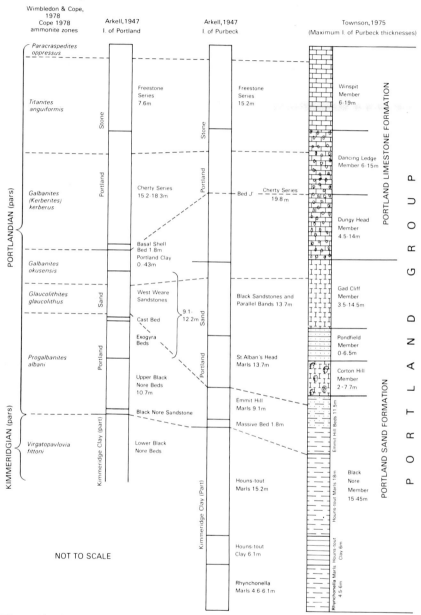

**Figure 12** Comparative classifications of the Portland Group in Dorset (After Arkell, 1947, p. 120, and Townson, 1975)

**Plate 6** Fossils from the Corallian Beds and Kimmeridge Clay
1 *Pavlovia rotunda*. 2 *Laevaptychus latus*. 3 *Lingula ovalis*. 4 *Thamnasteria concinna*. 5 *Thecosmilia annularis*. 6 *Hemicidaris intermedia*, *a* apical view, *b* side view. 7 *Nanogyra virgula*. 8 *Nanogyra nana*. 9 *Saccocoma sp*. 10 *Chlamys (Radulopecten) fibrosa*, *a* left valve, *b* right valve. 11 *Aulacostephanus autissiodorensis*. 12 *Torquirhynchia inconstans*, *a* dorsal view, *b* anterior view. 13 *Deltoideum delta*.

46  The Hampshire Basin

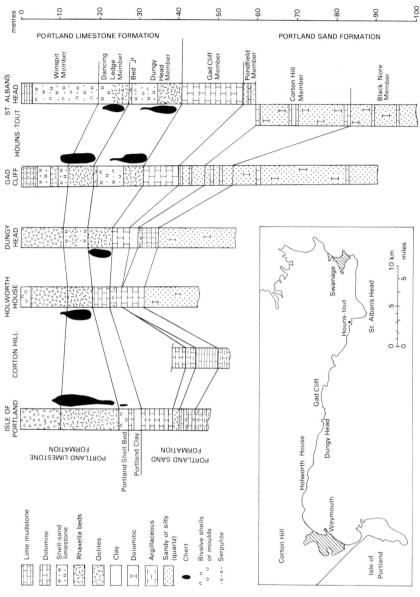

**Figure 13**
Correlations and facies variations in the Portland Group in Dorset
(From Townson, 1975)

include the Kimmeridgian ammonites *Pavlovia* and *Virgatopavlovia* and the Portlandian *Progalbanites*, bivalves including *Oxytoma* and *Nanogyra*, with *Rhynchonella*, belemnites, echinoid radioles and a few calcified spicules of the siliceous sponge *Rhaxella*. The beds are bioturbated, but not to the point where all traces of the original laminated bedding are obliterated.

The succeeding Corton Hill Member shows little marked change in lithology from the beds below in the 'East Basin', except for the incoming of beds of pale, pure lime mudstone with abundant *Nanogyra* at the base of each bed. Lime mud is the main clay-grade component of the Member over the 'Swell', but here the beds are mainly sand-grade dolomite, having been deposited as sand-grade limestone and dolomitised during or soon after deposition. In places massive accumulations of *Nanogyra* form the whole rock; the shells are not in life position but have been disturbed by bioturbation or current action. In the 'West Basin' the rock is either a dolomite with calcified *Rhaxella* spicules, or (as in the road cutting through Corton Hill) a sandy limestone with a variety of bivalves, many coated with serpulids. These beds accumulated slowly in shallow water and mark the end of the first regressive cycle.

The Pondfield Member was deposited in deeper water. It consists of partially dolomitised calcareous claystones with thin bands or nodules of limestone and a fauna of bivalves and sporadic ammonites. It thins over the 'Swell' to 0 to 0.2 m at Osmington Hill and Ringstead Bay.

The Gad Cliff Member is typically a greyish brown fine-grained dolomite with small amounts of clay and quartz sand, though in the 'East Basin' quartz sand may make up to 50 per cent of the rock. (It may be remarked that the Portland Sand Formation in Dorset never contains more than 50 per cent quartz sand and that the 'West Weare Sandstones' of the Isle of Portland (Arkell, 1947) contain less than 1 per cent.) Some beds are bioturbated by large (up to 1 m) *Rhizocorallium*. Shelly fossils are rare, which suggests that conditions were unsuitable for benthonic fossils (see p.49). Over the 'Swell', however, shelly fossils are more common and include the large ammonite *Glaucolithites*, bivalves, and the serpulid *Glomerula gordialis*. The rock has had a complex history, for the dolomite has itself been replaced locally by calcite pseudomorphs forming black nodules. At Dungy Head some of these nodules contain crystals of celestite. This member terminates the Portland Sand Formation.

The Portland Limestone Formation forms impressive vertical cliffs from Durlston Head to St Alban's Head and farther west at Gad Cliff. It also forms the cliffs that wall the Isle of Portland (Plate 8.1), and decrease in height gently southwards with the dip. It is divided into three members.

The lower beds of the Dungy Head Member in the 'East Basin' and over the 'Swell' are pale, bioturbated limestones rich in calcified *Rhaxella* spicules and with masses of chert, either in regular horizontal bands of nodules or in continuous beds up to 0.5 m thick traceable for hundreds of metres along the cliffs. The upper beds are fine-grained shell-fragment limestones, some of them browner and more heavily bioturbated than the lower beds. Their distinctness is emphasised by stylolitic upper and lower surfaces, where subsequent pressure has caused the limestone to dissolve. Some layers (called serpulites) are crowded with *Glomerula gordialis* and others with bivalves.

Over the 'Swell' this member is only 4 m thick and consists of 2.7 m of lime mudstone followed by 1.3 m of white chalky limestone. Farther west, in the Isle of Portland, it passes into a featureless, brown and grey dolomitic (ankeritic) clay, the Portland Clay, which can be seen interbedded with limestones at Holworth House.

The top bed of the Dungy Head Member is one of the most distinctive units in the Portland Group on the coast. In the 'East Basin' it is 'Bed J$^1$', 'Prickle Bed or Puffin Ledge' of Arkell (1947), a much bioturbated, nodular, shelly, greyish brown limestone with some chert. In the cliffs around St Alban's Head it has a level top surface, and the parting between it and the bed above has been hollowed out by weathering to provide (formerly) nesting places for puffins. Large ammonites are common and are encrusted with *Nanogyra* and *Glomerula,* indicating slow deposition. Towards the 'Swell' the bed is thinner. Here the correlation of the top of Bed J$^1$ with the top of the Portland Shell Bed of the Isle of Portland and the 'West Basin', first published by Cope and Wimbledon (1973), is easily made. On the Isle of Portland the Portland Shell Bed is 2.5 to 3 m of bioturbated shelly limestone in three or four layers. It is packed with a profusion of bivalves, gastropods, ammonites, serpulids, echinoids and other forms (Cox, 1929). There is much bioturbation, and encrustation of the larger fossils shows that deposition was mostly slow; but the exquisite preservation of some entire bivalves in life position also shows that some layers accumulated rapidly. This bed marks the end of the second regressive cycle.

The final cycle begins with the Dancing Ledge Member. This consists in the 'East Basin' of fine-grained, grey *Rhaxella* limestones with massive layers of dark grey to black chert, some shell beds, and a serpulite. These pass up into fine-grained shell-sand limestones with less abundant *Rhaxella*. Over the 'Swell' the limestones become chalkier and the cherts blacker; both are full of *Rhaxella*. Locally chert is seen to replace *Thalassinoides*. In the 'West Basin' the member is mostly a white lime mudstone. On the Isle of Portland the upper beds are mostly current-laid shell banks up to 4 m thick, with large ammonites and detached blocks of the coral *Isastraea oblonga*. There is much grey chert, and calcified spicules of *Rhaxella* and other sponges are abundant.

In the Isle of Purbeck the 'Cherty Series' of Arkell (1947) corresponds to the Dungy Head and Dancing Ledge members, but in the Isle of Portland it corresponds to the Dancing Ledge Member alone. The amount of chert is proportional to the concentration of *Rhaxella* spicules, which alone provide an adequate source of silica for the chert.

The Winspit Member comprises the building stones for which the Portland Group in this area is famous. In the 'East Basin' it consists mainly of shell-sand limestone with occasional oolitic layers and with identifiable shells only in the middle and upper parts. In this area the Shrimp Bed at the top is a smooth lime-mudstone up to 3 m thick. It is named after the fragments of a callianassid crustacean that it contains. It yields an ammonite, *Paracraspedites*, which gives a basis for correlation with contemporaneous beds in Lincolnshire and Russia (Casey, 1973). Over the 'Swell' and in the Isle of Portland, the member is mostly oolite. In the 'West Basin', it is a soft, chalky lime-mudstone, useless for building, but formerly burnt for lime.

The Winspit Member has been quarried not only on the Isle of Portland,

**Plate 7** Fossils from the Portland and Purbeck Groups
1 *Viviparus cariniferus*. 2 *Corbula sp*. 3 *Neomiodon medius*. 4 *Camptonectes (Camptochlamys) lamellosus*. 5 *Protocardia dissimilis*. 6 *Liostrea distorta*. 7 *Laevitrigonia gibbosa*, right valve. 8 *L. gibbosa*, internal mould of left valve.

where it is still worked, but also from the cliffs between Durlston Head and St Alban's Head, in places from adits commonly called caves, e.g. the Tilly Whim Caves, but all are now abandoned. Inland in Purbeck there is a large quarry near Worth Matravers. The famous 'Roach'—a greyish oolitic limestone honeycombed with the hollow moulds and casts of shells—is best known as the top bed of the Portland Group in Portland, but in fact the lithology appears at several levels in the Winspit Member. It is now found more useful than formerly, mainly as polished slabs for exterior cladding. The terms 'Whitbed' and 'Basebed' were formerly used for two beds of building stone, but the quality of the stone varies laterally and good material may be found at more than two levels (see Hounsell, 1952). Nowadays the chief product of the quarries is reconstituted stone, made by crushing and recompressing good, poor, cherty and chert-free limestones indiscriminately.

Scattered fossils, including the giant ammonite *Titanites*, occur in the Winspit Member. The commonest and most widely distributed are bivalves, notably *Protocardia dissimilis, Camptonectes (Camptochlamys) lamellosus, Isognomon listeri, Pleuromya tellina, Laevitrigonia gibbosa* and *Myophorella spp.* (Plate 7). The two last named (known to the quarrymen as ''osses 'eads') and a gastropod, *Aptyxiella portlandica* (the 'Portland Screw'), are particularly common in the 'Roach'. Local growths of calcareous algae termed 'patch reefs', up to 2 m thick and 1 m across, also occur, and testify to the shallowness of the water, since daylight was essential to their growth. Knots of the serpulid worm *Glomerula gordialis* are found, but are less common than in the lower, cherty limestones, where they form entire beds (serpulites), and in the Gad Cliff Member.

Apart from the cliff sections already mentioned, the Portland Group is seen at Pondfield Cove, where it dips to the north in the steep middle limb of the Purbeck Fold. It reappears at Mupe Rocks and continues westwards, rising in level and with steepening northerly dips, forming the coast—breached at Lulworth Cove—as far as Dungy Head, where it is caught in a fault and is seen only discontinuously as far as Durdle Door. It next appears in a faulted inlier above Ringstead Bay, from where it extends with few interruptions along the northern limb of the Weymouth Anticline to Portesham. Near Poxwell, an elongated east–west pericline exposes crags of the Portland Sand Formation. This picturesque hollow more than 1.5 km long is known as Poxwell Circus. It was the site of an oil exploration well in 1937.

To the north of the Weymouth Anticline, the Portland Group disappears beneath the Chalk, but is seen again north of the Dorset Downs in the Vale of Wardour. Here the classification of the beds was revised by Wimbledon (1976) and the ammonite zones by Wimbledon and Cope (1978). The Portland Sand Formation comprises three members. At the base, the Wardour Member (over 10 m thick) consists mainly of clays and bioturbated siltstones with a few *Nanogyra* horizons and some other bivalves. A bed of rounded pebbles of black chert occurs 3.5 m below the top, but there is no palaeontological evidence to correlate this with the Upper Lydite Bed of Swindon and Aylesbury or with the Rotunda Nodules of the Dorset coast (p. 43). The Chicksgrove Limestone Member (2 to 4 m) consists of well-bedded, pink and grey, fine-grained limestones with abundant bivalves. Ammonites (*Glaucolithites*) also occur. The Tisbury Glauconitic Member (about 12 m)

1  Cliffs south of Blacknor, Isle of Portland, showing beds of Portland Group capped by Lulworth Formation. Height of cliffs about 85.3 m. (A.12236)

2  Coastal landslip of the Lower Greensand in cliffs north-west of Blackgang Chine, Isle of Wight. Chalk cliffs to the east of the Needles in the background. (A.12024)

Plate 8

1  Stair Hole, west of Lulworth Cove, showing beds of Lulworth Formation dipping steeply to the north in the middle limb of the Purbeck monocline, and folded in the 'Lulworth Crumple'. (A.11225)

2.  Fossil forest, east of Lulworth Cove. 'Burrs' of stromatolitic algal growths surrounding hollows that were trunks of trees. (A.12224)

Plate 9

corresponds with the Main Building Stones of Arkell (1933). The beds are quartz sands, varying in colour (from grey to green), in glauconite content, in the abundance of shell fragments, and in degree of cementation. The lower half is well bedded; in the upper half, trough cross-bedding and mud-draped symmetrical ripples are seen.

The Portland Limestone Formation in the Vale of Wardour comprises two members. The Wockley Micritic Member (8 to 11 m) corresponds to the Ragstone Beds and Chalky Series of earlier accounts. The basal beds ('Ragstones') are well-bedded, shelly limestones with interbedded marls. The remainder consists of poorly-bedded, fine-grained, chalky limestones with scattered bivalves. At the top is the Chilmark Oolite Member (about 7 m), seen only in the east of the Vale and thinning out to the west. It consists of cross-bedded oolites with a few shell bands which yield *Aptyxiella portlandica*. Stromatolites like those of the Purbeck Group are interbedded with oolites at Chilmark, but elsewhere the junction with the Purbeck Group is gradational.

The ammonite zoning of the Portland Group in the Vale of Wardour is shown below:

| | | |
|---|---|---|
| *anguiformis* | | Chilmark Oolite |
| | Wockley Micritic | Member   7 m |
| *kerberus* | Member *including* | |
| | Ragstone Beds | 8 – 11 m |
| *okusensis* | | |
| | Tisbury Glauconitic Member | 12 m |
| *glaucolithus* | | |
| | Chicksgrove Limestone Member | 4 m |
| | Wardour Member | 10 m + ? |
| *albani* | | |

Farther north, in the Vale of Pewsey, the Portland Group crops out near Potterne, Coulston and Crockwood, but the present exposures are very poor.

Townson (1975) envisaged the deposition of the Portland Group (Figure 14) as having begun in relatively deep water where, as regression of the sea began, evaporating conditions may have set in around the margins of the sea. The water there may thus have become enriched in magnesium, would then have been heavier than normal sea water, and could have flowed over the sea floor to the lowest parts of the basin. Below a postulated critical depth, where the water was too alkaline and too stagnant (and hence too poor in oxygen) to support life, lime mud would have been dolomitised upon or within the sea bed. These conditions prevailed over most of the area during the deposition of the greater part of the Portland Sand Formation. Local exceptions are the shelly limestones of the Corton Hill Member at Portesham, and the *Nanogyra*-rich beds of the Gad Cliff Member on the 'Swell'. Above the critical depth, though the bottom muds were still too fluid to support benthonic animals, *Rhaxella* could survive and dolomitisation no longer occurred.

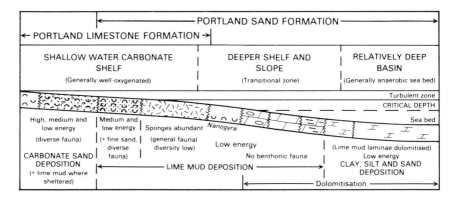

**Figure 14** General model for environments of deposition of the Portland Group of Dorset
(From Townson, 1975)

As the regression continued, so the water became shallower, better aerated and more turbulent. Food was more plentiful and the sea bed became more stable by the winnowing away of mud. Shell beds formed, some by growth in place, others by current drifting. In the shallowest and more agitated areas oolite banks formed (as over the 'Swell' from Lulworth to the Isle of Portland), while in protected lagoons behind them, algal stromatolites grew, isolating hypersaline areas in which evaporites were deposited, as seen in the succeeding Lulworth Formation. The diversity of possible environments in this phase is illustrated in Figure 15. Townson's sedimentological work has provided the first coherent picture of the deposition of the Portland Group and the first lithostratigraphic correlation between the Portland and Purbeck sequences.

**Lower part of the Purbeck Group**
The 'Purbeck Beds' (Purbeck Group of Townson, 1975) were divided into Lower, Middle and Upper parts and all included in the Jurassic until 1963, when Casey proposed to draw the Jurassic–Cretaceous boundary at the Cinder Bed of the Middle Purbeck. He called the Jurassic part the Lulworth Beds and the Cretaceous part the Durlston Beds (both classed as formations by Townson). He claimed that the Cinder Bed marked a marine transgression which he correlated with one known in the ammonite-bearing sequence in Norfolk and Lincolnshire and in Russia. Since no ammonite is known from the Purbeck Group, Casey's boundary is based on extrapolation from events in another basin of deposition. It is not now disputed, however, that the Purbeck Group straddles the Jurassic–Cretaceous boundary, and Casey's classification is followed here, without prejudice to arguments about the chronostratigraphical value of the Cinder Bed.

The older classification recognised the general lithological unity of the Purbeck Group, which is a sequence of alternating clays, marls and limestones, whereas the overlying Wealden is an alternation of clays, sands and sandstones. However, in places the topmost beds of the Purbeck Group are of Wealden appearance in that the clays become more brightly coloured

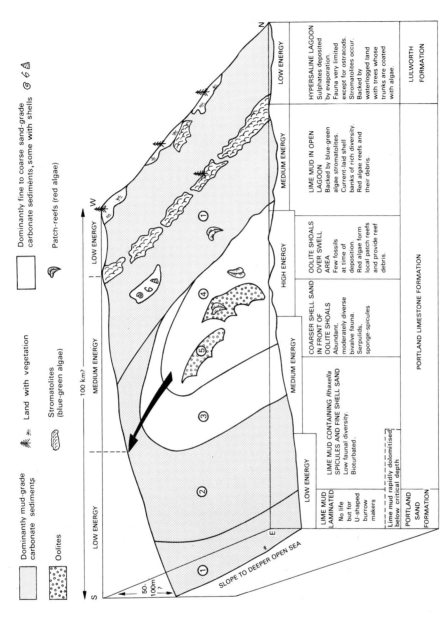

**Figure 15**
General model of maximum observed diversity of environments of deposition of the Portland Group of Dorset
(After Townson, 1975)

and the amount of sand increases. As the Purbeck – Wealden junction is nowhere well exposed, it is not certain that a consistent line has been followed in drawing the mapped boundary.

A drawing of the type-section in Durlston Bay is given by Clements (*in* Torrens, 1969). A horizontal section is shown in Figure 16.

**Lulworth Formation**

The sequence in the Lulworth Formation is shown in Table 6. The junction with the Portland Group is marked in some places by a fossil soil ('dirt bed'); in others there is a simple transition over half a metre or so from thick-bedded to thin-bedded limestones; in yet others, erosional channels cut into the Portland limestones. In all cases, however, massive, normal marine limestones of the Portland Group are eventually succeeded by laminated limestones representing restricted marine or lagoonal conditions.

Table 6  Thickness of the Lulworth Formation in Durlston Bay. (Measured by Dr R. G. Clements.)

|  | m | m | m |
|---|---|---|---|
| Cherty Freshwater Beds | 8.1 | | |
| Marly Freshwater Beds | | | |
| (to base of Mammal Bed) | 1.5 | | |
| 'Middle Purbeck Beds' part | | 9.6 | |
| Marly Freshwater Beds | | | |
| (below Mammal Bed) | 2.4 | | |
| Soft Cockle Beds | 24.8 | | |
| Hard Cockle Beds | 3.7 | | |
| Cypris Freestones | 13.7 | | |
| Broken Beds and Caps (after West, 1960) | 8.5 | | |
| 'Lower Purbeck Beds' Total | | 52.7 | |
| Lulworth Formation Total | | | 62.3 |

Evaporites occur everywhere at or near the base of the Lulworth Formation. In Sussex these are a major mineral resource of gypsum and anhydrite. In Dorset, by contrast, they have been replaced by other minerals (chiefly chalcedony and calcite). The complex processes involved have been worked out by I. M. West in a series of papers. In 1975 he summarised his own work on the sediments and that of others on the algae. He recognised four main facies, separated by dirt beds, as follows (Figure 17):

A  The lowest beds are limestones with foraminifera, calcispheres, ostracods, and brackish-water gastropods and bivalves with the isopod *Archaeoniscus*, but fossils are not abundant. Stromatolithic remains of blue-green algae of *Spongiostroma* type occur as separate algal mats or as confluent hummocky structures without regular lamination (these were regarded as tufas by earlier writers). This facies represents shallow lagoons and intertidal flats with moderately hypersaline water (50 to 70 parts per thousand)—not too salty for algal growth, but enough to discourage browsing molluscs.

B  Similar limestones with stromatolites, some of which are evenly laminated. The only fossils are a few ostracods. Many of the limestones are made up of small pellets of microcrystalline carbonate, and true oolites occur at Lulworth Cove, where the

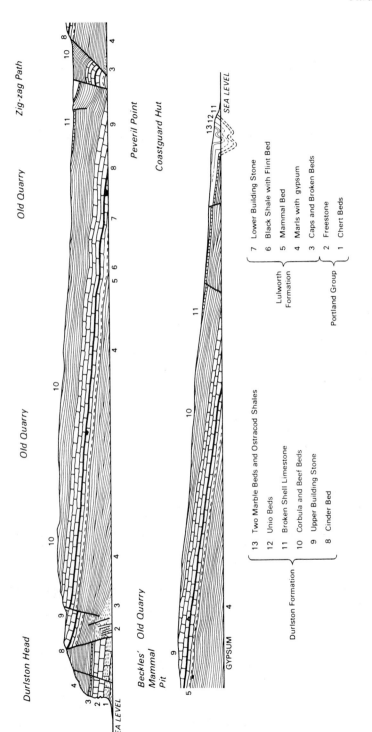

**Figure 16** Section from Durlston Head northwards to Peveril Point, Swanage. Distance 1.6 km. (After Strahan, 1898, p. 92)

58  The Hampshire Basin

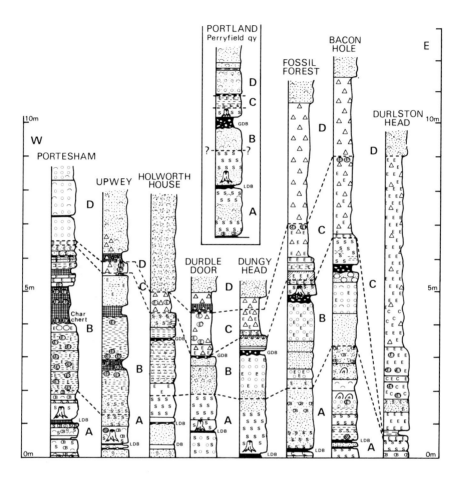

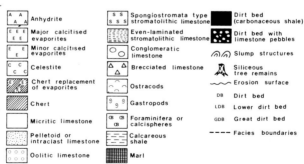

**Figure 17**  Stratigraphy and lithology of the basal Purbeck Group (Lulworth Formation) of the Dorset mainland
(From West, I.M. *in* Torrens, (editor), 1969, fig. A34)

**Plate 10** Some characteristic ostracods from the Purbeck Group. (Stereoscan photographs)
1 *Cypridea dunkeri* x 60. 2 *Fabanella boloniensis* x 33. 3 *Cypridea granulosa* subsp. *fasciculata* x 42. 4 *Cypridea granulosa* subsp. *granulosa* x 42. 5 *Theriosynoecum forbesi* x 62. 6 *Cypridea wolburgi* x 33. 7 *Cypridea tumescens* subsp. *praecursor* x 49. 8 *Cypridea vidrana* x 33.

famous Fossil Forest occurs in this facies (Plate 9.2). Small crystals of gypsum replaced by chalcedony or calcite are common, and the salinity must have approached the point—124 parts per thousand—at which gypsum is precipitated. The facies includes the Great Dirt Bed between Worbarrow Tout and Portesham, which represents a long interval of emergence.

C  Unfossiliferous limestones consisting almost entirely of calcite replacing anhydrite, which was itself secondary to gypsum, with some chert nodules replacing the sulphates. Thin layers of shale and of pelletoid limestone occur. Beds in this facies (which corresponds roughly with the Broken Beds, see Table 6) were deposited mainly as gypsum. They are well exposed at Durlston Head and Worbarrow Tout. The facies represents mainly supratidal lagoons in which gypsum was readily precipitated.

D  This facies differs from Facies A in that the pelletoid limestones are coarser and ripple-marked, and their lamination is due to small-scale current bedding. Stromatolites are not usually present but ostracods are abundant (the Cypris Freestones are in this facies). The salinity of the water was thus too high for molluscs but not for ostracods, and waves and currents were too energetic to allow algal mats to grow.

Some of the dirt beds contain limestone brash from the bed beneath, showing that lithification had been rapid. The Great Dirt Bed at Portesham contains land and freshwater shells and plant remains, as well as a famous fallen tree trunk 4.6 m long sheathed in a cylinder of stromatolite. The algae must have grown to that thickness before the tree fell, so that subsidence and deposition after the episode of the Great Dirt Bed must have been rapid.

It was for long debated whether the Broken Beds were brecciated by collapse after solution of the evaporites, or by tectonic action. In the lumps of Facies-C limestone, however, the evaporites had already been replaced before brecciation, and Arkell (1947) gave other reasons for concluding that the brecciation was due to tectonic movement.

The Cypris Freestones are followed by the Hard and Soft Cockle Beds, which show that the area was submerged and that deposition was controlled by cyclic, perhaps climatic, influences. Brackish-water episodes, with molluscs such as *Protocardia, Neomiodon* and *Hydrobia* and such ostracods as *Cypridea* (Plate 10), are represented by limestones which thicken westwards, towards the basin margin. They show that influxes of fresh water brought down large volumes of sediment (with remains of many kinds of insects) and diluted the stagnant, hypersaline waters that periodically covered tidal flats. Here grew mats of the algae *Ortonella* and *Girvanella* which, with the surrounding mud, cracked on desiccation. Many bedding planes are covered by crushed remains of *Archaeoniscus*.

The highest beds of the Lulworth Formation consist mainly of shales and calcareous clays (Marly Freshwater Beds) succeeded by hard cherty limestones in which the fossils are silicified and beautifully preserved (Cherty Freshwater Beds or Lower Building Stones). At the base of these there formerly occurred a remarkable lenticular bed of marl (the Mammal Bed) which yielded a large number of bones of mammals, similar in size to rats and cats; young crocodiles (originally thought to be dwarf adults); and over 150 specimens of lizards. The bed seems to have been entirely dug away.

Only the lower beds of the Lulworth Formation remain on the Isle of Portland (Plate 8.1), where nearly 30.5 m are exposed in the cliffs north and west of Southwell. Elsewhere the beds have been quarried away. The succession is similar to that in Purbeck.

The Lulworth Formation thins westwards, at first slowly (from 62.3 m at Durlston Bay to 52.1 m at Worbarrow Bay and 47.5 m at Mupes Bay) and then more quickly. The margin of deposition may have been not far west of Portesham.

In the Vale of Wardour the Lulworth Formation begins with 'caps' and dirt beds, as in Dorset. Many well-preserved fishes have been found, probably in these beds. The thickness has been variably estimated and may change over short distances; it probably does not much exceed 24 m. Away from the outcrop it is 36.3 m at Kingsclere and 47.6 m at Portsdown.

# 4. Cretaceous rocks

### Upper part of the Purbeck Group (= Durlston Formation) and Wealden

Above the Lower Building Stones is a bed which indicates the most considerable change in conditions in Purbeck times, namely, an invasion of the sea which spread over a large part of the south of England. This marine episode left behind it conspicuous beds made up almost entirely of oystershells. From their appearance where seen in a weathered condition these are known as the Cinder Beds in Durlston Bay; they are present in the Vale of Wardour as well as in south Dorset. In addition to the oysters (*Liostrea distorta*) they include shells such as *Laevitrigonia, Isognomon* [*Perna*], *Myrene* and *Protocardia*, while the occurrence of *Hemicidaris purbeckensis*, a small seaurchin, is of especial interest. In the Vale of Wardour the bivalves *Laevitrigonia* and *Myrene* outnumber the oysters. In the Vale of Pewsey the bed is a reddish sandstone which extends as far north as Buckinghamshire (Whitchurch Sands). Casey and Bristow (1964) recorded a fauna that increasingly resembles a normal marine assemblage as the bed is traced northwards; and they suggested that the sea which deposited ammonite-bearing latest Jurassic and earliest Cretaceous strata in Lincolnshire and Norfolk may have broken through to the south at this time. It is this transgression which Casey (1963) took to mark the beginning of the Cretaceous. The general Cretaceous succession is shown in Table 7.

The succeeding Intermarine Beds or Upper Building Stones, Scallop Beds, Corbula Beds and Chief Beef Beds (Table 8) are limestones, some of them sandy, interbedded with shales. The fauna varies from normal marine (*Chlamys* in the Scallop Beds) to marine, though suggesting slightly reduced salinity (*Protocardia, Corbula*, etc.), and brackish to freshwater conditions (*Neomiodon, Hydrobia, Viviparus*). The building stones have been extensively mined and quarried in the Isle of Purbeck. They have yielded a rich variety of reptile footprints, mostly of dinosaurs, which indicate that the beds were exposed while still wet—perhaps between tides, perhaps at times of low water level in the Purbeck lagoon. These have been thoroughly reviewed by Delair (1958 – 60). Freshwater conditions prevailed almost without interruption towards the close of Purbeck times, for the Upper Purbeck beds all show evidence of having been so deposited. At the base is the Broken Shell Limestone or Burr, composed to a large extent of shell fragments, and attaining a thickness of 2.9 m. It has been used as a building stone. Above are clays with layers of beef and bands of greenish limestone with shells of the freshwater mussel *Unio*. These are the Unio Beds. The higher part of the Upper Purbeck is important as yielding the famous Purbeck Marble. This material has been used for interior decoration since Roman times and may be seen in many churches and cathedrals. It occurs in two layers, red and green, and is made up to a large extent of the univalve shell *Viviparus* (formerly known as *Paludina*). The Marble Beds are included in the Upper Cypris Clays

Cretaceous 63

**Table 7** Cretaceous succession in the Isle of Wight and Dorset

| GEOLOGICAL AGE | | ENVIRONMENT Non-marine / Marine Shallower ← retreating / Deeper → advancing | FORMATIONS | MAIN TYPES OF ROCK | Cumulative thickness in metres | Theoretical age in millions of years |
|---|---|---|---|---|---|---|
| UPPER CRETACEOUS | Coniacian to Campanian | Highest beds missing in this region | UPPER CHALK | Chalk with flint | 100 / 200 / 300 | 65 / 70 / 75 / 80 |
| | Turo-nian | | MIDDLE CHALK | Chalk with some flint | 400 | 85 |
| | Ceno-man-ian | | LOWER CHALK | Muddy chalk on sandy phosphatic basement bed | 500 | |
| LOWER CRETACEOUS | Albian | | UPPER GREENSAND AND GAULT | Shallow marine sands with phosphate / Deeper marine clays | 600 | 90 |
| | Aptian | | LOWER GREENSAND | Shallow marine sands and clays | 700 / 800 | 95 / 100 |
| | Ryazanian to Barremian | | WEALDEN | Well bedded shales with shelly layers; brackish to marine at top / Brightly coloured marls with (in Dorset) cross-bedded sandstones and coarse grits | 900 / 1000 / 1100 / 1200 / 1300 / 1400 / 1500 | 105 / 110 / 115 / 120 / 125 / 130 |
| | | | DURLSTON BEDS | Non-marine limestones and clays, marine at base | | 135 |

**Table 8** Thickness of the Durlston Formation in Durlston Bay. (Measured by Dr R. G. Clements.)

|  | m | m | m |
|---|---|---|---|
| Upper 'Cypris' Clays and Shales (partly obscured) | 12.9 | | |
| Unio Beds (average) | 1.1 | | |
| Broken Shell Limestone | 2.9 | | |
| 'Upper Purbeck Beds' Total | | 16.9 | |
| Chief Beef Beds | 8.3 | | |
| Corbula Beds | 11.4 | | |
| Scallop Beds | 1.4 | | |
| Intermarine Beds | 15.5 | | |
| Cinder Bed | 2.9 | | |
| 'Middle Purbeck Beds' part | | 39.5 | |
| Durlston Formation, Total | | | 56.4 |

and Shales. At the top are the Viviparus Clays (0.1 to 3.4 m thick), not seen at Swanage, which consist of coloured clays and marls of variable thickness, and contain abundant shells of *Viviparus*. These clays are exposed at Worbarrow Bay and Mupe Bay.

In the area north of Weymouth the Purbeck beds have been quarried for building stone near Upwey. The formation crops out again in the Vale of Wardour, but is thinner; the tiny ostracods show, however, that both Jurassic and Cretaceous divisions are represented. The beds have been quarried at several places between Dinton and Tisbury.

According to H. W. Bristow's measurements (1857), the Durlston Formation thins westwards more rapidly than does the underlying Lulworth Formation: from 56.4 m in Durlston Bay, to 36.6 m at Worbarrow Bay, 29.6 m at Mupes Bay and 25.6 m at Ridgeway, north of Weymouth. These measurements may, however, be treated with some scepticism, because the Durlston Formation passes conformably and gradually up into the clays and sands of the Wealden. In Dorset, where the best exposures of the transition occur, 'no line can be drawn which does not either include beds of Purbeck type in the Wealden or beds of Wealden type in the Purbeck, the two formations being absolutely inseparable' (Strahan, 1898). Any line of division is thus arbitrary; and there is no way of correlating it accurately from one section to another. With these reservations, thicknesses of 16.5 and 25 m are reported in the deep boreholes at Kingsclere and Portsdown.

The Durlston Formation in the Vale of Wardour is probably about 23 m thick. At the base, the Cinder Bed is a rough grey sandy limestone. The beds corresponding to the Upper Purbeck have never been fully exposed, but they include clays and sands which 'appear to indicate . . . . what may be termed Wealden conditions . . . . This is only what might be expected to occur on the margins of the Purbeck basin'—placed by Andrews and Jukes-Browne (1894) within 32 km to the north.

The Wealden, in spite of magnificent coast sections in the Isle of Wight (where 305 m out of a total thickness of 612 m are exposed) and in Dorset, is probably the least studied formation in the region. The gradual transition from the generally clayey and limy sediments of the Durlston Formation to the sandy and pebbly beds of at least the lower part of the Wealden may mark a gradual change in climate because it was accompanied by a revival of active rivers. These rivers, draining a land area in the west, spread their detritus over southern England and northern France in the form of extensive delta-flats with braided channels. The deposits are of varied character, including coloured sands, sandstones, grits, mottled clays and shales. Plant-debris beds occur and some tree trunks and rolled bones are found. Oldham (1976), from a study of the plant cuticles in the plant-debris beds, envisaged their deposition on a delta top or delta front and shoreline. He compared the vegetation with that found today in southern Florida. In the Purbeck area the Wealden beds include an abundance of quartz pebbles and quartz grit, which was presumably brought from the west by river action; but some of the heavy minerals indicate derivation from an ancient massif extending from northwest Spain to Cornubia which comprised a range of igneous and metamorphic rocks of various ages (Allen, 1972).

When the volume of sediment decreased through the lowering of the surrounding land masses, large freshwater lagoons were formed, and the deposits of these waters gave rise to finely laminated shales. For the most part, sedimentation kept pace with subsidence, but occasional periods of emergence, when the deposits became dry enough for large reptiles to roam over them, are indicated by the presence of suncracks or footprints. One interesting feature, indicative of the origin of these beds, is the 'Pine Raft', once visible at low tide near Brook Chine, in the Isle of Wight. This was a mass of prostrate tree trunks that were evidently washed into the delta, where they became waterlogged and stranded.

In the Isle of Wight, where the upper part of the Wealden is brought to the surface by two anticlines, two distinct divisions can be recognised (Figure 18). The lower division comprises the Wealden Marls or Variegated Clays (167 m seen); the upper is the Wealden Shales (58 m). The Marls are red, purple, green and variegated clays and marls, with bands of sandstone, sand and sandy limestones; they contain much driftwood, ferns, fruits of conifers and shells (among them the freshwater snail *Viviparus spp.* and the freshwater mussel *Margaritifera (Pseudunio) valdensis* (Plate 11). Fish remains and water-worn bones of large reptiles (*Iguanodon*) are also to be found. The Shales are dull blue or blackish clays with layers of clay ironstone, sandstone and shelly limestone; their even bedding and dull hue present a marked contrast to the coloured and massive Marls. Some of the shale layers are so finely bedded as to constitute 'paper shales'; their surfaces are commonly crowded with the remains of ostracods (Plate 10), which enable the beds to be correlated with the upper part (*Cypridea valdensis* Zone) of the Weald Clay of the Weald (Anderson, 1967). Near the base of the Shales in Brixton Bay is a red, sandy bed which has yielded remains of the small dinosaur *Hypsilophodon* (fully redescribed by Galton, 1974). Higher up, and conspicuous in the cliffs, is a massive bed, the sandstone of Barnes High. Among the larger fossils of the Shales, in addition to the shells mentioned as occurring in the Marls, are the bivalve *Filosina gregaria* and sundry species of ferns. In the upper beds bivalves (oysters and the cockle *Nemocardium ibbetsoni*) and the brackish-water gastropod *Cassiope fittoni* are common.

The Wealden Marls are of freshwater origin, but in the highest beds of the Shales there is evidence of the proximity of marine conditions. At the end of

66  *The Hampshire Basin*

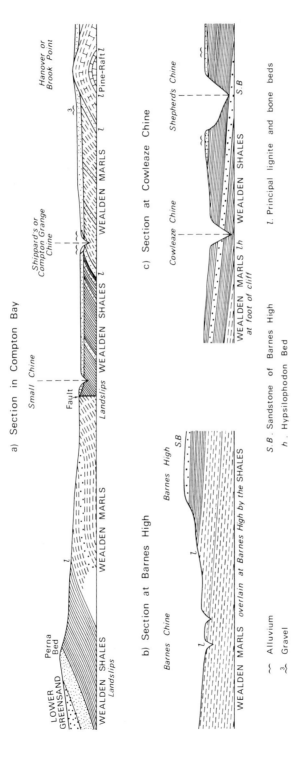

**Figure 18**  Sections of the Wealden Beds in the Isle of Wight
Horizontal scale 1:8000: vertical scale exaggerated. (After White, 1921, p. 12)

Wealden times earth movements that had been in progress became more appreciable and eventually the sea, advancing from the south, broke over most of the area of the Wealden delta. It occupied the south of England as far west as the centre of Dorset and extended northwards through Wiltshire, laying down deposits called the Lower Greensand. The line of junction between the two formations is sharply defined, and it is evident that some of the Wealden beds were eroded by the advancing sea because at the base of the Lower Greensand are fossils from the Wealden and even older beds.

The Wealden beds show a thickness of 716 m at Swanage; thence to the west they become coarser and conglomeratic, and decrease rapidly in thickness, being 366 m in Worbarrow Bay, 229 m at Mupe Bay, and 107 m at Upwey, near Weymouth. Their further extension westwards was eroded during the deposition of the Lower Greensand or the Gault. The beds are well exposed on the Dorset coast in the five great sections at Swanage, Worbarrow and Mupe bays, at Lulworth Cove and Durdle Promontory. In the Isle of Wight about 305 m are exposed (about 61 m of Wealden Shales) but 612 m were proved in a borehole at Arreton. They are 252 m thick at Portsdown, 341 m in boreholes near Winchester, and 282 m at Kingsclere. In the Vale of Wardour about 2 m of reddish sand at Dinton may belong either to the Wealden or to the marine Whitchurch Sands. In a large intermediate area around Wareham, Fordingbridge and Blandford they are absent.

## Lower Greensand

The Lower Greensand marks the first full development, following preliminary signs in the topmost beds of the Wealden, of a great cycle of marine sedimentation that lasted until the end of the Cretaceous. The deposits are seldom green and many are not sand; and the name seems to have arisen from mere confusion with the green and sandy deposits above the Gault Clay—the Upper Greensand.

The stratigraphy of the Lower Greensand was thoroughly revised and its fauna listed by Casey (1961), who is also monographing the ammonites of the formation (1960 – ). In the Isle of Wight and in Swanage Bay the junction with the underlying Wealden Shales is sharp and marked by a thin gritty and pebbly layer which indicates some interruption in deposition.

The fullest development of the Lower Greensand in the area is in the south of the Isle of Wight, where (near Atherfield) it attains a thickness of 246 m (Figure 19). It decreases in thickness eastwards, westwards and northwards, being 183 m in the eastern end of the island and 122 m in the west, while at Punfield, on the Dorset coast, it is less than 61 m; it does not reach 15 m in the Vale of Wardour. In Dorset the formation is not found farther west than Mupe Bay, where its thickness is 20 m, except for a 0.1-m fossiliferous ironstone on the east side of Lulworth Cove which caps sands that pass down without a break into the Wealden.

Altogether 16 'Groups', numbered from I to XVI in Table 9, have been recognised near Atherfield, but for general purposes the following four-fold division is used:

68  *The Hampshire Basin*

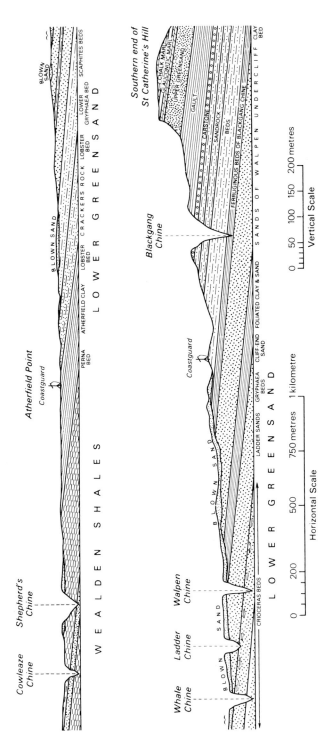

**Figure 19** Cliff section of Lower Greensand, from Atherfield Point to St Catherine's Hill, Isle of Wight (After White, 1921, p. 26)

**Table 9** Stratigraphy of the Lower Greensand at Atherfield, Isle of Wight. (Vertical scale 1:1000)

| Stage | Zone | Bed | No. | Formation |
|---|---|---|---|---|
| Albian (Lower) | *Douvilleiceras mammillatum* | Carstone | XVI | Carstone |
| | *Leymeriella tardefurcata* | Sandrock | XV | Sandrock |
| Aptian (Upper) | *Hypacanthoplites jacobi* | | | |
| | *Parahoplites nutfieldensis* | Ferruginous Bands of Blackgang Chine | XIV | Ferruginous Sands |
| | | Sands of Walden Undercliff | XIII | |
| | *Cheloniceras martinioides* | Foliated Clay and Sand | XII | |
| | | Cliff-End Sand | XI | |
| | | Upper Gryphaea Beds | X | |
| | | Walpen and Ladder Sands | IX | |
| | | Upper Crioceras Beds | VIII | |
| Aptian (Lower) | *Tropaeum bowerbanki* | Walpen Clay and Sand | VII | |
| | | Lower Crioceras Bed | VI | |
| | *Deshayesites deshayesi* | Scaphites Beds | V | |
| | | Lower Gryphaea Bed | IV | |
| | *Deshayesites forbesi* | Upper Lobster Beds, Crackers, and Lower Lobster Bed | III | Atherfield Clay |
| | | Atherfield Clay | II | |
| | *Prodeshayesites fissicostatus* | Perna Bed | I | |

Carstone
Sandrock Series
Ferruginous Sands
Atherfield Clay (with the Perna Bed at the base)

The two upper divisions are not normally fossiliferous. In the lower divisions some species, such as the brachiopod *Sellithyris* (Plate 11) and the oyster *Aetostreon [Exogyra] latissimum*, range throughout, but otherwise many of the beds have rich and characteristic faunas.

Groups I to III compose the Atherfield Clay in the broad sense. Group I, the Perna Bed (1.5 m), is named after a large, coarsely-ribbed bivalve, *Mulletia* (formerly *Perna*) *mulleti*. The basal gritty layer contains derived Jurassic ammonites (*Pavlovia*) and derived and indigenous fish teeth. Its top layer is a richly fossiliferous sandstone yielding the coral *Holocystis elegans*, *Mulletia* and other large bivalves, nautiloids and many other fossils. Ammonites are very rare.

Group II, the Atherfield Clay proper (18.3 to 21.3 m), consists of brown-weathering, greyish silty clay, which wastes away so fast that clear sections are rarely visible. Bivalves and gastropods provide the commonest fossils. Ammonites are less common and occur either as crushed impressions or as clay-ironstone internal moulds of the body-chamber. Bunches of branched tubes composed of little pellets (*Granularia*) may be of faecal origin or represent burrow fillings.

Group III, about 26 m thick, consists of the Lower and Upper Lobster Beds with the Crackers—two lines of large doggers in 6 m of firm clayey sand—between. In the Lower Lobster Bed, fossils are abundant and well preserved, usually in thin clusters, though fossil prawns (*Meyeria magna*) are commoner than true lobsters. *Deshayesites* and *Roloboceras* are the dominant ammonites. The rich and varied bivalve and gastropod fauna is like that of the Crackers. These doggers, which form a feature in the cliff about 550 m E of Atherfield Coastguard Station, are named from the explosion of air pockets trapped by the waves in cavities hollowed out beneath them. They have yielded a rich profusion and variety of well-preserved fossils: echinoderms, bivalves, gastropods, ammonites (mainly *Deshayesites*), the peculiar belemnite phragmocone *Conoteuthis*, crabs, cirripedes and fishes. The Upper Lobster Bed is only sporadically fossiliferous.

The Ferruginous Sands (Groups IV to XIV) begin with the Lower Gryphaea Bed (8.5 m), named from the massive *Aetostreon latissimum* which was once mistaken for a *Gryphaea*. The fauna has much in common with that of the Crackers, but the species of *Deshayesites*—in particular *D. deshayesi*—are different. Leaf impressions of the fern *Weichselia* also occur. The succeeding Scaphites Beds (V; 15.2 m thick), named from the ancyloceratid ammonite now called *Australiceras (Proaustraliceras)*, is again widely fossiliferous and mark the highest occurrence of *Deshayesites*. *Cheloniceras* is also common.

The Lower Crioceras Bed (VI) is 4.9 m thick and is exposed at the mouth of Whale Chine. It contains small, intensely hard, phosphatic concretions packed with fossils. The giant crioceratid *Tropaeum* and the ammonite *Dufrenoyia* first appear here. Similar ammonites, with many bivalves, occur in Group VII, the Walpen Clay and Sand (17.37 m), which extends as far as Ladder Chine.

The Upper Crioceras Beds (VIII) are about 14 m thick. They are best studied up the slopes of Whale Chine. Fossils tend to occur in clusters, and include characteristic species of *Cheloniceras (Epicheloniceras)* and other ammonites with many bivalves. The greyish brown clayey sand is impregnated with calcite in many places and contains drift wood and fronds of *Weichselia*. The Walpen and Ladder Sands (Group IX) are 12.8 m of greenish grey sand with a line of large calcareous concretions at the base. These, which are about 0.45 m thick and 1.2 m long, each contain a brown phosphatic core full of fossils. Calcite lines every crack and cavity and fills the interiors of shells. Many of the cores are nests of young ammonites (*Cheloniceras*) or of the long-ranging brachiopod *Sellithyris*, but other ammonites and a variety of bivalves and other fossils also occur. The Upper Gryphaea Beds (X; 4.9 m), consist of ferruginous clayey sands which yield little but *Aetostreon latissimum*. From here upwards the Lower Greensand is relatively less fossiliferous except at certain levels to be noted.

The 8.5-m of the Cliff-End Sand (XI) yield poorly preserved *Cheloniceras*. In the upper part there are remarkable, cylindrical, branching concretions. No fossils have been found in the Foliated Clay and Sand of Group XII (10.6 m).

The yellow and brown, glauconitic Sands of Walpen Undercliff (XIII) are about 27.4 m thick. They extend for about 640 m under Blackgang Chine. Near their middle and to the east of the chine occur large fossiliferous nodules yielding immature *Parahoplites*, bivalves and other fossils. Similar, but less weathered nodules at Horse Ledge, Shanklin, yield a more varied ammonite fauna.

The Ferruginous Bands of Blackgang Chine (XIV) consist of about 6.1 m of sands with three lines of irony concretions. These yield a profusion of bivalves and gastropods, but no ammonites. Cycads have also been recorded. The overlying 12.2 m of dark grey sandy clay are unfossiliferous.

The Sandrock (XV) consists of 56.7 m of unfossiliferous, white and yellow sand and sandrock. The Carstone (XVI) comprises 3.7 m of gritty and pebbly sands with phosphatic nodules and reaches a thickness of 22 m at Red Cliff, north of Sandown. In Reeth Bay, Puckaster Cove and Watershoot Bay, either side of St Catherine's Point, these nodules have yielded many ammonites—*Sonneratia, Douvilleiceras*, etc.—bivalves, gastropods and echinoids.

Dike (1972) interpreted the trace fossil *Ophiomorpha* (the burrow-fillings of callianassid Crustacea) in the top beds of the Ferruginous Sands and in the Sandrock Series as marking a change from mainly offshore marine sedimentation to shoreline and nearshore conditions.

Between Shanklin and Sandown the succession is difficult to correlate in detail with the Atherfield sequence just described. Three horizons are of special interest to palaeontologists. One is the Urchin Bed, forming the southern part of Horse Ledge at Shanklin Point. Apart from nine species of echinoids, this bed yields a variety of branching polyzoa, brachiopods, bivalves and gastropods. Fragmentary ammonites (*Parahoplites*) suggest a correlation with Group XIII. Secondly, the succeeding 6.1 m contain the well-known ferruginous concretions of Shanklin—masses of bivalves and gastropods preserved as moulds and tightly packed together. These correlate with the Ferruginous Bands of Blackgang Chine (XIV). Thirdly, at the mouth of Luccomb Chine may be seen a bed of grit 2.4 m thick, some 6.1 m above the base of the Sandrock. Nodules at the base of this grit yield a rich flora. They are the original source of the Benettitales, an extinct group of cycad-like plants whose structure suggests an abortive attempt to evolve into angiosperms. The beds also contain many conifers and possibly angiosperms as well.

At Red Cliff, although the whole of the Lower Greensand (here 183 m) is exposed, the beds are deeply weathered and sparsely fossiliferous. The Perna Bed and Atherfield Clay can, however, be seen in a large fallen mass north of Yaverland Fort, and are richly fossiliferous as at Atherfield.

At Compton Bay, in the western part of the island, the Lower Greensand is reduced to 122 m and the beds have changed so much in character that close correlation with the Atherfield section is impossible.

On the Dorset coast, the Lower Greensand is 60.5 m thick at Punfield Cove, at the northern end of Swanage Bay. Here, above a 0.9-m basal Pebble Bed, are 14.17 m of Atherfield Clay capped by the 0.3-m Punfield Marine Band, which is followed by 45 m of practically unfossiliferous Ferruginous Sands. The Atherfield Clay yields the bivalves of the *forbesi* Zone and, near the top, the characteristic *Deshayesites* of the Lower Lobster Bed. The Punfield Marine Band, correlated with the Crackers, contains rare ammonites of that horizon and many fossils suggesting brackish-water conditions: *Cassiope*, rare in the Crackers at Atherfield; and the bivalves *Nemocardium ibbetsoni*, found in both the Wealden Shales and the Atherfield Clay, and *Eomiodon*, unknown in the Crackers. Farther west, at Corfe, at Worbarrow Bay and finally at Lulworth Cove, the fauna becomes more brackish, in transition towards a repetition of the Wealden facies. These facts were interpreted by Casey (1961) as indicating a slight elevation of the land which caused the deposition of the Atherfield Clay at Atherfield to be interrupted by the sandy episode of the Crackers. As this horizon is traced westwards the fauna becomes less and less marine, as would be found in travelling up the estuary of a river.

Inland the Lower Greensand is exposed in small isolated areas. A narrow outcrop south of Shaftesbury extends for about 10 km along the border of the Gault. The thickness here does not exceed 12 m and the formation comprises beds of sand, clay, and glauconitic and sandy clays, with occasional mottling. In the absence of fossils it is not possible to refer these beds to any particular subdivision.

In the Vale of Wardour the Lower Greensand is represented by 4.5 to 6 m of glauconitic sand with traces of chert, but the outcrop is unimportant and exposures scarce. A wider outcrop with a few outliers extends west and north-west of Devizes, where the formation includes ferruginous sands, sandy loams and ironstones. A small outlier caps the hill at Seend, and there the beds include an ironstone that was formerly quarried, the ore being smelted on the

**Plate 11** Fossils from the Wealden Group and the Lower Greensand
1 *Margaritifera (Pseudunio) valdensis*. 2 *Sulcirhynchia hythensis*, *a* dorsal view, *b* anterior view. 3 *Mulletia mulleti*. 4 *Deshayesites wiltshirei*. 5 *Sellithyris sella*, *a* dorsal view, *b* anterior view. 6 *Meyeria magna*. 7 *Australiceras gigas*. 8 *Tessarolax fittoni*.

spot. The beds here have yielded a large number of fossils, including the ammonite *Parahoplites nutfieldiensis* and many shells of brachiopods and molluscs.

Farther north, near Calne, is another small outcrop, of up to 12 m of ferruginous sands and sandstones with subordinate clay seams, bands of pebbles and a few lenses of white sand. The fossils are mostly casts and moulds of bivalved shells and brachiopods, among them the bivalve *Toucasia lonsdalei*, one of the rare English representatives of the mainly palaeotropical rudists.

These inland outcrops of the Lower Greensand show clearly that the formation in this part of the country was laid down by a transgressive sea. Earth movements had brought certain of the Upper Jurassic strata within the range of marine erosion. As a result, in different parts of this inland area, the Portland, Purbeck and Wealden beds have been planed away before or during the transgression of the sea. In various localities, therefore, the Lower Greensand rests successively on the Wealden Beds, the Purbeck Beds, the Portland Beds and the Kimmeridge Clay (as in the Vale of Wardour).

## Gault, Upper Greensand, and basal beds of the Chalk

A further downward movement followed the deposition of the Lower Greensand, a movement more rapid and extensive than that which submerged the area over which the Wealden beds were deposited. There was also slight general tilting towards the east and uplift towards the west, so that the sea, as it swept westwards eroding as it went, spread its deposits on successively older strata; these deposits (the Gault and Upper Greensand) overstep all the Jurassic formations until they rest on the Triassic, beyond the western borders of our area. This can be seen when the outcrop is traced from Swanage, where the succession beneath the Gault is nearly complete, to near Lyme Regis, where the Gault and Upper Greensand rest successively on Middle Lias (as on Golden Cap), the Middle and Lower Lias junction (as on Stonebarrow) and the Lower Lias (as on Black Ven).

The Gault and Upper Greensand are successive Albian formations of different lithology. The former name is used where the deposits are mostly clays; the latter where sandy beds predominate; but the fossils prove that clay beds in one part of the country are of the same age as sandy developments elsewhere. In general terms the sandy element becomes more conspicuous as the formation is traced westwards, until near Lyme Regis the Gault is not only thin but sandy, while the Upper Greensand is much more prominent. Fossils of all groups are found in these formations; and species of ammonites, by reason of their restricted vertical range, provide a means of subdivision into zones. By means of the ammonites, also, beds in different parts of the country can be correlated irrespective of their lithic character. The relationship of the Upper Greensand to the basal beds of the Chalk is so complex as to make a combined description more helpful than separate ones.

In the Isle of Wight the Carstone passes up through silty micaceous clays to the Gault clay, which is about 30 m thick at Red Cliff and about 20 m in Compton Bay (Owen, 1971). The Gault clay passes up in turn through sandy clays and marls 4.5 to 13.4 m thick into the Upper Greensand. This formation is from 24.4 to 27.4 m thick throughout the island. It is composed mainly of speckled, light greenish to bluish grey sandstones. Small chocolate-coloured concretions occur in most of the beds. At certain horizons the strata abound in hard phosphatic nodules and on weathered surfaces these give a characteristic rugged appearance. Some beds contain large concretions or doggers, while the higher beds are characterised by chert in scattered concretions and in regular courses. The Upper Greensand gives rise to some conspicuous features in the Isle of Wight, such as the cliff which dominates the Undercliff from Bonchurch to Blackgang, and the inland bluffs of the Southern Downs. In the central range it forms the scarp of Rams Down and hog-backed ridges south of the Chalk near Brading.

On the Dorset coast the Gault, with a pebble bed at its base, oversteps the Lower Greensand at Lulworth Cove and transgresses over successively older formations to come to rest on Lower Lias on Black Ven. Its thickness fluctuates somewhat as it is traced from Punfield Cove (27.7 m) to Black Ven (7.3 m).

Inland from the coast the Gault, here reduced from 3 to 6 m of bluish grey sandy clay, is mostly hidden by swamps and landslips. At Okeford Fitzpaine 8.3 m were formerly exposed, and near Ansty some 10.6 m have been recorded. From here it thickens northwards to over 27 m in the vales of Wardour and Pewsey. Boreholes show that it is 52 m thick beneath the Chalk in the upper Kennet Valley and 86.5 m at Kingsclere.

Ammonite evidence for the zoning of the Gault is poor over the whole area. It seems mostly to be of early Middle Albian age, but Owen (1971) suggested that its upper part may be of Upper Albian age in the Isle of Wight.

The outcrop of the Upper Greensand and basal beds of the Chalk follows roughly that of the Gault, zigzagging in a narrow strip round the major structures of the Hampshire Basin syncline, the anticlines of the vales of Wardour and Pewsey, and the syncline between them. The geology of these beds is, however, much more difficult to understand on a regional basis than that of the Gault. This is because the upper beds of the Upper Greensand vary from place to place both in lithology and in age (as reflected in the ammonite faunas), while the basal beds of the Chalk too are of different ages in different places (Figures 20 and 21).

For the purpose of describing the stratigraphy, the Upper Greensand outcrop may be divided into four areas, defined below. The strata are of Upper Albian age in most of the region, but in the second and the fourth of the four areas they include beds of Cenomanian age. In the second area the succession is much reduced, and in the fourth it is more complete. There is nearly everywhere a visible non-sequence at the base of the Chalk and the basal beds become progressively younger as they are traced south-westwards from a line joining the Vale of Wardour to the Isle of Wight. This age change is not, however, a simple or regular process.

The first area comprises the Isle of Wight and the Dorset coast as far west as Holworth House. Here the Gault clay is separated from the Upper Greensand (and, at Holworth House, replaced) by bluish sandy clays termed 'Passage Beds'. They have yielded no diagnostic fossils. The Upper Greensand (about 20 to 30 m thick) consists mainly of glauconitic, ill-sorted sands and sandstones. Many of the beds contain irregular lumps or larger, spheroidal doggers ('Cowstones') of calcareous sandstone and layers of phosphatic nodules. A 7-m member near the top of the formation in the Isle of Wight is named the

76   The Hampshire Basin

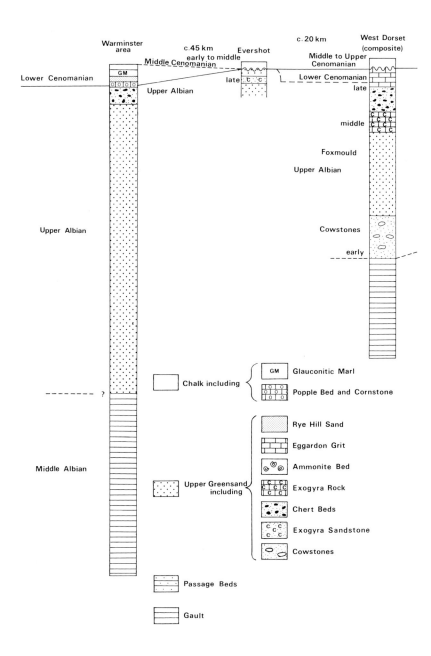

**Figure 20**   Relationships of Albian and Cenomanian strata in Dorset and the Isle of Wight
The terms 'early', 'middle' and 'late' are used in an informal sense. Vertical scale 1:200. (For distribution of sections, see Figure 21)

# Cretaceous

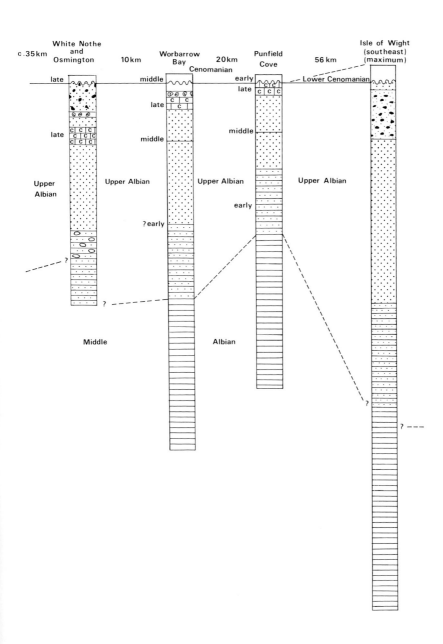

78  *The Hampshire Basin*

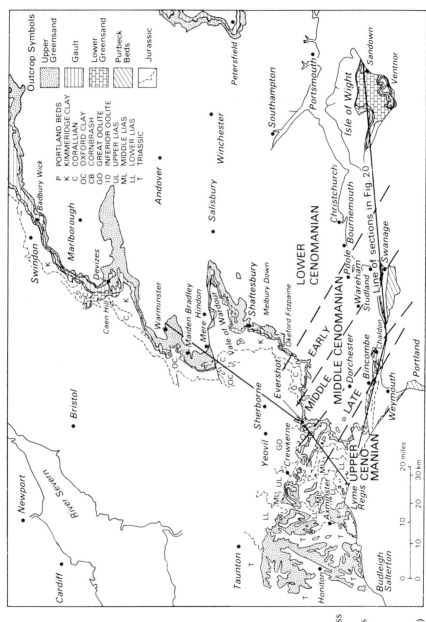

**Figure 21**
Diagram of
Cenomanian
transgression across
Dorset
(Note line of sections
in Figure 20)
(Modified from
Owen, 1971, fig. 18,
with information
from Kennedy, 1970)

Chert Beds. At approximately the same position in east Dorset is the Ammonite Bed, rich in phosphatised ammonites of latest Albian age.

The top surface of the Upper Greensand is everywhere uneven and is in many places coated with a veneer of phosphate. It may be pitted and fissured to a depth of 10 cm or more, with material from the base of the Chalk filling the cavities. The basal bed of the Chalk, the Glauconitic (formerly 'Chloritic') Marl (up to 2 m) is, however, of early Lower Cenomanian age, so that the time interval was probably too short to measure by the methods of stratigraphy.

Where the base of the Chalk is next seen, at Punfield Cove, it is on the north-eastern fringe of the area in which deposition throughout late Albian and Cenomanian time was influenced by a buried feature named the mid-Dorset Swell (Hancock, 1969; Drummond, 1970; Kennedy, 1970). The effect of this feature was progressively to delay the onset of Chalk sedimentation. Thus at Punfield Cove the Chalk Basement Bed (so named because it is younger than the Glauconitic Marl, though similar to it in appearance) is of early Middle Cenomanian age; at Worbarrow Bay it is of middle, and at Holworth House of late Middle Cenomanian age. The age of the bed is determined by the latest, and least phosphatised, ammonites that it yields; but it may well also contain earlier Cenomanian and derived upper Albian ammonites, of various degrees of phosphatisation.

The main Cretaceous outcrop swings inland north-westwards from Holworth House through Portesham to Litton Cheney; from there north to Hooke; then in a north-westerly tongue towards Crewkerne, before being cut off by the east–west Chelborough Fault belt which shifts the outcrop east to beyond Buckland Newton. This area, together with the south-western outliers between Beaminster and Charmouth, forms the second area of Upper Greensand lithology. Unfortunately, except for the coastal outliers, little is exposed apart from the beds a few metres below and above the base of the Chalk.

On the coast (Golden Cap to Black Ven) the lower part of the formation begins with sands with Cowstones (about 7 m), followed by yellow, iron-stained sands known as Foxmould. The commonest fossil here is *Amphidonte obliquata* (Plate 12), formerly *E. conica*), and in places this occurs in masses with other oysters forming the Exogyra Rock (of middle Upper Albian age—not the same as the late Upper Albian Exogyra Sandstone of east Dorset). This is succeeded by Chert Beds (up to 4 m), and this in turn by the Eggardon Grit (up to 3 m), with the intervention near Maiden Newton of another layer of Exogyra Rock. The Eggardon Grit has yielded latest Albian ammonites near Rampisham and Batcombe; but at Eggardon Hill Lower Cenomanian ammonites are found in its upper part. In this area, therefore, the top beds of the Upper Greensand are of about the same age as the lowest beds of the Chalk in the Isle of Wight.

In this area, as in the first, the top of the Upper Greensand is uneven and phosphatised; and the Chalk Basement Bed is of Middle or, at the western limits of the region, of Upper Cenomanian age.

The third area of Upper Greensand extends east and north from near Buckland Newton to near Shillingstone. Here, above dark green sands and loams of unknown thickness, the top of the Upper Greensand consists of a nodular, glauconitic, rubbly calcareous sandstone with dark brown phosphate nodules. In places the top part of this bed has clearly been broken up in Albian times and reworked into pebbles, some of which lie in fissures and hollows in the lower part of the bed. The fissuring is dramatic in its effects at some localities. Lower and even Middle Cenomanian ammonites may be found as far as 30 cm below the top of the Upper Greensand. The Chalk Basement Bed in this area becomes progressively older north-eastwards until, at Shillingstone, it yields both phosphatised and unphosphatised early Lower Cenomanian ammonites.

The fourth area of the Upper Greensand is around Warminster and Mere (Figure 22). The account given here largely follows Jukes-Browne and Scanes (1901). At Rye Hill Farm, half a metre of glauconitic sand, the Rye Hill Sand, overlies the Cornstones which rest on about 3 m of Chert Beds. The Rye Hill Sand is one of the richest sources of Lower Cenomanian (so-called 'Warminster Greensand') fossils in England. This is the second of the two occurrences of Cenomanian Upper Greensand mentioned on p. 75.

At Mere and Maiden Bradley the succession is condensed. At Mere the Chert Beds are followed by the Cornstones, here called the Popple Bed (below). At Maiden Bradley the Chert Beds, with Upper Albian ammonites at the top, are followed by Cornstones, succeeded by brown glauconitic sands with phosphatic nodules (not directly correlated with the Rye Hill Sands), followed by the same succession as at Mere.

The Popple Bed is a complex mixture of pebbles ('popples') of sandstone derived from the bed beneath, from which it is separated by an irregular erosion surface. It is followed by normal Glauconitic Marl, and the two together yield a fauna of early to middle Lower Cenomanian age. The Popple Bed is here taken as the basal bed of the Chalk because it cannot be distinguished from the Glauconitic Marl above.

From older records it is known that the Upper Greensand is about 40 m thick in the Warminster area. So far as is known this thickness is approached, but not exceeded, where the formation is concealed beneath Chalk to the east.

Laing (1975) has divided the beds from the upper part of the Weald Clay to the top of the Lower Chalk into four zones based on angiosperm pollen, using material from the Isle of Wight and Dorset. The usefulness of this work will be found in correlating non-marine sediments of this age with the standard ammonite-zoned marine succession.

**Plate 12**  Fossils from the Gault, Upper Greensand and Lower Chalk
1 *Mortoniceras inflatum*. 2 *Neithea gibbosa*. 3 *Mantelliceras mantelli*, *a* lateral view, *b* ventral view. 4 *Merklinia aspera*. 5 *Inoceramus (Birostrina) concentricus*. 6 *I. (Birostrina) sulcatus*. 7 *Holaster subglobosus*, *a* apical view, *b* side view. 8 *Orbirhynchia mantelliana*, *a* dorsal view, *b* anterior view. 9 *'Cardiaster' fossarius*, *a* apical view, *b* side view. 10 *Amphidonte obliquata*. 11 *'Rotularia'* cf. *umbonata*. 12 *Stoliczkaia dispar*. 13 *Turrilites costatus*. 14 *T. acutus*.

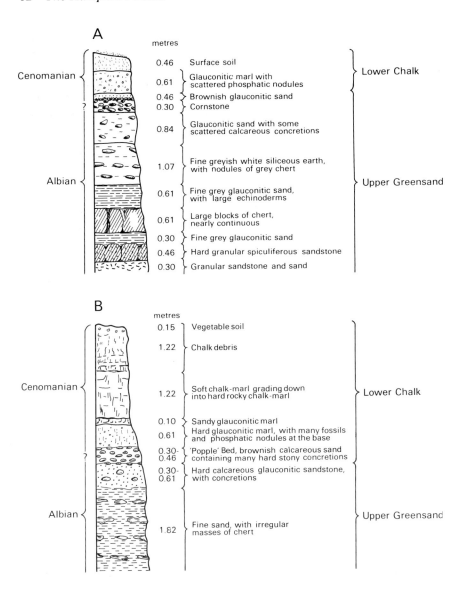

**Figure 22** Upper Greensand and Lower Chalk in sections near Warminster
A Maiden Bradley Quarry and B Dead-Maid Quarry, Mere. (Adapted from Jukes-Browne and Scanes, 1901, pp. 101, 111)

## The Chalk

The constitution, and the mode and conditions of deposition, of the Chalk were reviewed by Kennedy and Garrison (1975). Normal soft chalk is made up of whole or broken coccoliths (the calcite plates which form the armour of microscopic planktonic algae—the coccolithophorids), up to 10 μm in diameter. These form a matrix in which coarser calcite components—calcispheres (*Oligostegina*), foraminifera, and fragmentary debris of many larger invertebrates (especially the bivalve *Inoceramus*)—are embedded. Such chalk was deposited in water of 200 to 300 m depth (below the photic zone), and practically undisturbed for most of the time, to allow the extremely fine sediment to settle. The top 0.1 m or more of the bottom was a fluid carbonate 'mud' containing 50 to 60 per cent water; this passed down into a thixotropic layer (one that grows firmer if undisturbed, but that liquifies if disturbed, for example by currents or burrowing animals), and that in turn into permanent sediment undergoing lithification by the precipitation of calcite from the interstitial water. The whole of the Chalk is extensively bioturbated, but by a limited range of creatures: among the traces of these, *Thalassinoides* predominates; *Chondrites* and *Planolites* (the back-filled burrows, up to 15 mm diameter, of sediment-eaters, probably worms) are also common.

The Chalk sea was an open shelf sea of much wider extent than any Jurassic sea in our area, and corresponding with the outer part of a continental shelf. The sediment shows clear evidence of cyclic deposition, with episodes of low energy and steady accumulation alternating with ones of higher energy and perhaps shallower water, when deposition was interrupted. The causes of these cyclic changes are not understood. There is no change in the trace-fossils through these cycles (except that pre-existing burrows may have been used by different creatures boring into lithified layers). But the main benthonic fauna changed from a narrow range of forms specially adapted to life on a fluid substrate (*Inoceramus*, some *Spondylus*, '*Cidaris*', etc.; these provided settlement areas for a wider range of creatures) to a more varied suite of animals taking advantage of a firmer bottom. Thus (in contrast with, for example, the cyclothems of the Lias) there was little or no change in such ecological factors as temperature, salinity and oxygenation.

It is clear that the thixotropic Chalk muds, when left undisturbed, were rapidly lithified (perhaps in a few tens of years). In some cases, the shallow, zero-deposition phases of cycles represent water of only 50 to 100 m depth, and in these cases complex processes leading to the formation of simple or composite hardgrounds (described in detail by Kennedy and Garrison, 1975) were initiated. The first sign that these processes had begun is an 'omission surface', which may be marked by a faint colour change, by a band of nodular flints (with processes protruding into burrows in the sea floor), or by a seam of marl, or more obvious evidence of non-sequence—the formation of nodules, their cementation or coalescence into a continuous bed, erosion of soft chalk, encrusting of the sea floor, and borings into consolidated chalk, especially by clionid sponges. Glauconite and phosphate may also have been developed. Some omission surfaces may be recognisable over more than 1000 km$^2$. They are only weakly discernible in areas of thick deposition, but where the succession is thin and discontinuous, they are unmistakable. Gritty chalks, which are simply normal chalks from which the coccolith mud has been gently winnowed, may be seen towards the top of Chalk cyclothems.

Flints formed at several different periods during and after the deposition of the Chalk. The earliest were those that replaced burrow fillings, thus outlining omission surfaces that are otherwise indistinct. At a later stage these may have been joined together by further growth of flint to form more or less continuous courses of nodules; these were nevertheless formed before the beginning of Tertiary deposition in England, for they appear as pebbles in Palaeocene and later gravels. Latest of all, and perhaps in part of post-Cretaceous age up to the present day, are sheets of tabular flint lining inclined or vertical joints and fault-planes in the Chalk.

In this region, Chalk deposition began either with the Glauconitic Marl or with the complex Chalk Basement Bed, both described above. The stratigraphic classification of the succeeding Lower, Middle and Upper Chalk is shown in Table 10. Wright and Kennedy (1981) have provided a sequence of ammonite zones for the Plenus Marls and Middle Chalk and have co-ordinated them with the traditional zones.

**Table 10** Stratigraphic classification of the Chalk of the region following Rawson and others, 1978

| Stage | Zone | 'Division' |
|---|---|---|
| Campanian | Belemnitella mucronata<br>Gonioteuthis quadrata<br>Offaster pilula | Upper Chalk |
| Santonian | Marsupites testudinarius<br>Uintacrinus socialis<br>Micraster coranguinum<br>(upper part) | |
| Coniacian | M. coranguinum<br>(lower part)<br>Micraster cortestudinarium | |
| Turonian | Holaster planus<br>Terebratulina lata<br>Mytiloides labiatus | Middle Chalk |
| Cenomanian | Sciponoceras gracile<br>Calycoceras naviculare<br>Acanthoceras rhotomagense<br>Mantelliceras mantelli | Lower Chalk |

**Lower Chalk**

Above the Glauconitic Marl or Chalk Basement Bed the main mass of Lower Chalk consists of the Chalk Marl and the Grey Chalk. Both divisions show alternations of more and less marly layers. The Chalk Marl varies in thickness up to 30 m, and is mostly a bluish grey or buff marly chalk. It is characterised by ammonites of the genera *Schloenbachia*, *Mantelliceras* and *Hypoturrilites*, bivalved shells including *'Aequipecten' beaveri*, *Entolium sp.*, *Plicatula inflata* and *Inoceramus crippsi*, and tubes of the worm *'Serpula'*

1 Cliffs of Durdle Cove, Dorset, showing a vertical bed of Chalk eroded by wave action along a thrust-plane. (A.12229)

2 Dry valley in Chalk country near Warminster, Wiltshire. (A.9708)

**Plate 13**

*umbonata*. In the Isle of Wight the sponge *Exanthesis labrosus* is particularly common near the base. Occasionally grey flinty lumps with a core of black flint are found in the Chalk Marl. Carter and Hart (1977) recognised a mid-Cenomanian non-sequence marked by an increase in the ratio of planktonic to benthonic foraminifera in the microfauna. The Grey Chalk, which in places attains a thickness of more than 30 m, consists of poorly fossiliferous pale grey or yellowish white slightly marly chalk, massively bedded. Its fossils include the oyster *Pycnodonte*, and other bivalved shells such as *Inoceramus pictus*, the echinoids *Holaster subglobosus* (Plate 12) and *H. trecensis*, and tubes of *'Terebella' lewesiensis*, a burrowing organism whose burrow was lined with detached fish-scales. Ammonites of the genus *Acanthoceras* are found in the lower part of this division, as are also *Turrilites* and *Scaphites*. In the Vale of Pewsey there is a bed 4.25 m thick of soft, whitish siliceous chalk with numerous concretions. The concretions are flinty or cherty nodules of irregular shape, bluish grey when broken, and are quite different from ordinary chalk flints.

The upper limit of the Lower Chalk is marked by the Plenus Marls (0.6 to 1.8 m), comprising greenish grey marly bands with layers of lighter chalk and conspicuous omission surfaces. The upper part of the marls is characterised by the belemnite *Actinocamax plenus*.

## Middle Chalk

The Middle Chalk varies in thickness up to 60 m, although not much more than half that amount in places. The beds near the base have a distinctly nodular structure, being hard and white in places and elsewhere containing slightly yellow nodules and lumps in pale grey streaky marl. Over most of the area a definite basal rock-bed is thus recognised—the Melbourn Rock (0.3 to 4.27 m)—but this is not distinct in south Dorset and the Isle of Wight. Above, the chalk becomes firm, white and homogeneous, with marl partings here and there. Flints occur as scattered nodules or in layers. There are no flints in the Middle Chalk of the Isle of Wight, but near the top of this division there are two bands of siliceous nodules which look like flints but have the texture of chalk. At Swanage and in the Isle of Wight, the upper part of the MiddleChalk becomes lumpy and nodular and a conspicuous hardground of yellowish white chalk with green-coated nodules results. This band is called the 'Spurious Chalk Rock'.

The commonest fossils of the Middle Chalk are brachiopod shells, echinoids and bivalve shells. Two small species of brachiopod are characteristic; one, a globose, plainly-ribbed form (*Orbirhynchia cuvieri*, Plate 14) is typical of the basal part of the Middle Chalk; the other, a much smaller and flatter form with fine divergent ribs (*Terebratulina lata*) is common at some horizons in the *lata* Zone. Among the echinoids in the lower zone

**Plate 14** Fossils from the Chalk
1 *Micraster coranguinum*. 2 *M. praecursor*. 3 *Mytiloides mytiloides*. 4 *Micraster cortestudinarium*. 5 *Discoides dixoni*, *a* apical view, *b* side view, *c* oral view. 6 *Marsupites testudinarius*. 7 *Uintacrinus socialis*, *a* external, *b* internal views of calycal plate, *c* brachial plate. 8 *Offaster pilula planatus*, *a* apical view, *b* side view. 9 *'Cidaris' hirudo*. 10 *Acanthoceras rhotomagense*, *a* side view, *b* ventral view. 11 *Orbirhynchia cuvieri*, *a* dorsal view, *b* anterior view. 12 *Belemnitella mucronata*.

are *Conulus castanea, Discoides dixoni,* and *Hemiaster (Peroniaster) nasutulus* (formerly known as *Hemiaster minimus*) and in the upper zone *Holaster planus* and *Micraster sp.* Two common species of bivalve shells may be noted; *Mytiloides mytiloides,* restricted to the lower part, and forms of the *Inoceramus lamarcki* group, characteristic of the upper part.

**Upper Chalk**
Most of the Chalk in the area belongs to the Upper Chalk, which attains a thickness of 404 m in the Isle of Wight and up to 260 m in Dorset. In many places its base is marked by the best-known hardground complex in the Chalk, the Chalk Rock. This bed is normally conspicuous on account of the presence of green-coated chalk pebbles. The Chalk Rock contains a peculiar assemblage of fossils, including sponges and hollow moulds of ammonites and gastropods. The Chalk above this bed becomes smooth, white and massively bedded, except for nodular beds and hardgrounds in the *cortestudinarium* and *coranguinum* zones in the Isle of Wight and the *pilula* Zone of the Isle of Wight and Dorset. Flints occur throughout, in nodular form, sometimes set in regular courses or in tabular seams. Thin seams of marl are also present at certain horizons. No chalk is now preserved above the lower part of the *mucronata* Zone.

Most of the principal divisions of the animal kingdom are presented among the fossil species of the Chalk. Remains of fishes occur usually as isolated teeth, those of the sharks *Cretilamna* and *Scapanorhynchus* and crushing teeth of the ray *Ptychodus* being the most common. Among the echinoids, *Micraster* (Plate 14) provides the zone fossils of the lower part of the Upper Chalk, and shape variants of *Echinocorys* characterise particular horizons throughout the Upper Chalk. Ammonites are frequent in the lower part of the Chalk and have been useful for purposes of correlation. In the Middle and Upper Chalk the few specimens found outside the Chalk Rock are of interest for long-distance, but not for local, correlation. Gastropods also are found in the lowest beds, and among the genera represented are *Aporrhais, 'Pleurotomaria', Semisolarium* and trochids. Bivalve shells are found throughout the Chalk: oysters, pectinids and *Spondylus* are well represented, while *Inoceramus* and its allies are of zonal value. Brachiopod shells tend to be concentrated at particular horizons; *Orbirhynchia* ranges throughout, *Gibbithyris* ranges up to the *coranguinum* Zone and *Cretirhynchia* throughout the Upper Chalk. Among the polyzoa, *Bicavea rotaformis* is notable as occurring in profusion at the base of the Upper Chalk in the Isle of Wight in a bed (1.9 to 2.4 m) called the 'Bicavea Bed'. Sponges, already mentioned from the lowest beds, are frequent in the Upper Chalk, being commonly enclosed in flint.

On account of its thickness and fairly uniform constitution in a comparatively simple tectonic setting the Chalk formation gives rise to extensive areas of characteristic scenery—that of the Downland type (Plate 13.2), with the associated escarpments and coombes. Irregular lines of escarpments lie along the western boundary of the Chalk. The slopes and height of the escarpments depend on the inclination of the strata, being bold and high where the dips are slight. Thus in the Vale of Wardour the escarpment on the south is prominent, but on the northern side, where the dips are much greater, the escarpment is dominated by the ridge formed by the Upper Greensand.

Behind the escarpments are wide areas of undulating upland country, which, where bare of superficial deposits, form open downs covered with short turf but generally without trees. Where the Chalk is covered by superficial deposits such as brickearth or clay-with-flints, tracts of woodland occur, but these have been cleared over wide areas.

The uplands are dissected by irregular and branching systems of valleys with steeply sloping sides. Such valleys are dry but they have the appearance of having been formed by running brooks and streams. They were formed under conditions which no longer prevail, when the water-table in the Chalk was higher than at present.

# 5. Tertiary (Palaeogene) rocks

## Introduction

By the end of Cretaceous time the sea had retreated completely from the region. The Chalk was gently folded and exposed to erosion so that the earliest Tertiary rocks rest on different horizons of the Chalk at different places; in the area there is no equivalent of the Thanet Beds of the London Basin, so that it is nearly everywhere the Reading Beds that are involved. In the Isle of Wight and at Studland Bay these beds rest on various horizons in the *mucronata* Zone (see Table 10). Elsewhere older beds of the Chalk were exposed, and in the Vale of Pewsey the Reading Beds overstep the whole of the Upper Chalk to rest on Middle Chalk.

The surface on which the Tertiary rocks were laid down was nevertheless one of low elevation and low relief. The open sea occasionally returned to cover the whole area, but only for short periods. For most of the time of which we have visible record in the rocks, the area was covered by shallow inshore seas, or by deltas, estuaries and lagoons on the eastern and southern margins of a low-lying hinterland. Changes in the facies of the deposits were rapid over short distances and in some cases over wide areas. Their interpretation is debatable because they represent changes in a number of different factors and processes—relative sea level, base level of erosion, temperature, salinity and degree of aeration of water, distribution of currents, sources of sediment, and climate. Keen (1977) gave a new analysis of some of these changes, based on a study of the assemblages of ostracods in the Headon, Osborne and Bembridge Beds. A generalised and speculative attempt to represent gross environmental changes is made in Figure 23.

Blondeau and Pomerol (1969) postulated three main source-areas for the Tertiary sediments of the Hampshire Basin, each providing a different suite of heavy minerals. The Cornubian hinterland provided a steady supply of minerals common to practically all the beds—zircon, tourmaline and rutile; a northern source, whose influence is particularly evident in the more marine sediments, supplied hornblende, epidote, zoisite and garnet; an Armorican source in the area of the Cherbourg peninsula, the Channel Islands and Brittany provided metamorphic minerals, notably kyanite and staurolite. The more continental deposits are in some cases almost devoid of heavy minerals, which suggests repeated panning and reworking of the sediment; when they yield heavy minerals, these indicate a strong Armorican influence. Clay minerals that swell on wetting, collectively termed smectite (which includes montmorillonite), are commoner in the more marine sediments. Kaolinite and illite are more common in the terrestrial sediments. Gilkes (1978) thought that the clay minerals of the Upper Eocene and Oligocene beds indicated that they were mainly formed of reworked Lower Eocene material from the margins of the basin of deposition.

The Tertiary rocks of the Hampshire Basin are separated from those of the London Basin by the anticlinorium of the Weald and by the Chalk of the

Tertiary (Palaeogene) 91

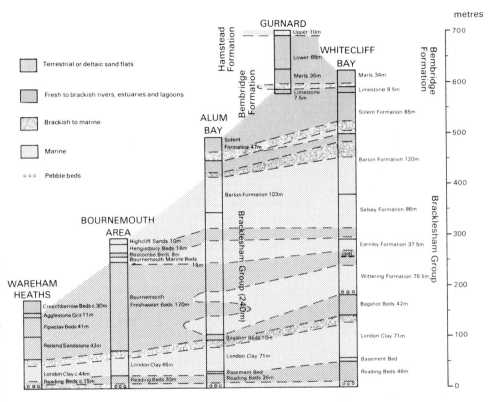

**Figure 23** Generalised succession of facies in the Palaeogene Beds of the Hampshire Basin
(Modified from Curry, 1965)

Hampshire Downs (Figure 24). The two basins are, however, structural basins, defined by tectonic movement. Both formed part of a single basin of deposition, the Anglo-Paris-Belgian Basin. Samples from the floor of the English Channel (Curry and Smith, 1975) allow a Hampshire–Dieppe Basin to be defined within the larger basin. This is bounded on the south-west by an important structural line running from Bembridge (Isle of Wight) towards Dieppe. Within the Hampshire–Dieppe Basin the succession is thickest and most complete in the eastern part of the Isle of Wight. Beneath the Channel, all the Eocene sediments are marine, but the southern part of the basin was probably land in Oligocene times, with large areas of exposed Chalk. These may have been the source of the fresh flints found in pebble beds in late Eocene formations in the Isle of Wight.

Westwards, beyond a landward continuation of the Bembridge–Dieppe line, those formations that persist beyond Christchurch and Wimborne become increasingly sandy and pebbly, and yield few fossils useful in correlation. Possibly in Middle Eocene times and perhaps later, that part of the basin outside—that is, north and west—of a line running approximately from near Woking to near Bournemouth was being filled with sandy deposits, ranging

92  *The Hampshire Basin*

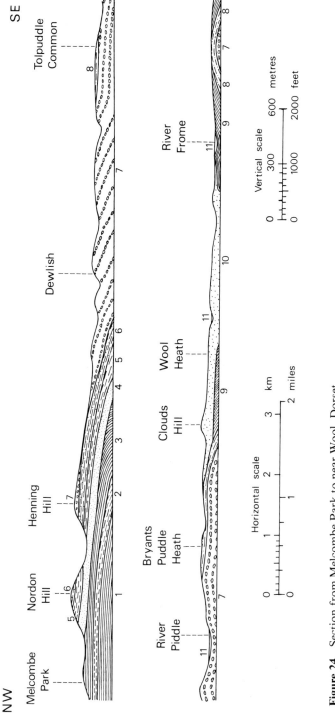

**Figure 24** Section from Melcombe Park to near Wool, Dorset
(After Whitaker and Edwards, 1926, fig. 2)

1. Oxford Clay
2. Corallian Beds
3. Kimmeridge Clay
4. Gault and Upper Greensand
5. Lower Chalk
6. Middle Chalk
7. Upper Chalk
8. Reading Beds
9. London Clay
10. Bagshot Beds
11. Alluvium

from fully terrestrial heaths and dunes to deltas and sand flats cut by braided channels. Inside—that is, south and east—of the line were large areas covered more or less completely and continuously by water—fresh, brackish, or marine. The sediments formed in this latter area yield ample fossil evidence for dating and correlation, and happen also to be the best exposed; but correlations into the peripheral sandy belt are hazardous in the light of current knowledge.

In Eocene times the climate, as reflected in the land flora, was that of tropical lowlands such as those of south-east Asia today, though certain bivalves such as *Astarte* and *Arctica* suggest that cooler waters circulated in the sea from time to time. At the beginning of Oligocene times a marked cooling set in and the flora indicates subtropical swamp conditions like those of present-day Florida. There is floral evidence suggestive of a cooler, drier and more elevated distant hinterland.

The classification of the Palaeogene rocks shown in Table 11 is adapted from Stinton (1975, pp. 6-7) and Cooper (1976). The Eocene–Oligocene boundary has been placed as advocated by Cavelier (1976). An up-to-date summary of the Palaeogene rocks of the Isle of Wight was given by Hancock and Curry (1977).

Bujak and others (1980) have published a monograph on the dinoflagellates and acritarchs of the Eocene of southern England and propose the following zones for the London Clay, Bracklesham Group and Barton Clay (see Figure 26 and Table 11).

| Zone | | Stratigraphical position | Locality |
|---|---|---|---|
| BAR-5 | *Polysphaeridium congregatum* | 49 to 56 m above base of Barton Beds | Whitecliff Bay (not known at Barton) |
| BAR-4 | *Homotryblium variabile* | 34 to 49 m above base of Barton Beds Barton Beds H (part) to I (part) | Whitecliff Bay Barton |
| BAR-3 | *Rhombodinium porosum* | Barton Beds E to H (part) | Barton |
| BAR-2 | *Areosphaeridium fenestratum* | Barton Beds A3 (part) to D | Barton |
| BAR-1 | *Heteraulacocysta porosa* | Barton Beds A1 to A3 (part) | Barton |
| B-5 | *Cyclonephelium intricatum* | Bracklesham Group, Beds 19a to 19c of Eaton, 1976; XVII to XIX (part) of Fisher, 1862 | Whitecliff Bay |
| B-4 | *Areosphaeridium arcuatum* | Bracklesham Group, Beds 10b to 18 of Eaton, 1976; VII to XVI of Fisher, 1862 | Whitecliff Bay |
| B-3 | *Phthanoperidinium comatum* | Bracklesham Group, Beds 9a to 10a of Eaton, 1976; V (part) to VI of Fisher, 1862 | Whitecliff Bay |
| B-2 | *Pentadinium laticinctum* | Bracklesham Group, Beds 5 to 7 of Eaton, 1976; IV to V (part) of Fisher, 1862 | Whitecliff Bay |

| | | | |
|---|---|---|---|
| B-1 | *Homotryblium abbreviatum* | Bracklesham Group, Beds 1 to 4 of Eaton, 1976; I to III (part) of Fisher, 1862 | Whitecliff Bay |
| LC-3 | *Kisselovia reticulata* | | |
| LC-2 | *Membranilarnarcia ursulae* | London Clay, 43 to 90 m above base | Whitecliff Bay |
| LC-1 | *Deflandrea phosphoritica* | London Clay, 0 to 43m above base | Whitecliff Bay |

## Palaeocene

### Reading Beds

The Reading Beds, which are 50 m thick in Whitecliff Bay, retain elsewhere a fairly uniform thickness of about 32 m except where they are thinner in the extreme west of their outcrop. They rest on an eroded and piped Chalk surface. The basement bed is a layer of large unworn flints with black or dark green exteriors, embedded in about 3 m of loamy glauconitic sand. Westwards to about the longitude of Wimborne this bed may contain sharks' teeth and comminuted oyster shells. Farther west it becomes coarser, and gradually loses its glauconite and fossils. The flints are in places decomposed and associated with irregularly developed ironstone.

Above the Basement Bed the strata are very variable. They may consist almost entirely of brightly mottled clays, or almost entirely of sand with lenticular bodies of gravel. Layers of ironstone occur in places and well-preserved leaves have been found in the clays.

Permanent exposures of the Reading Beds are accessible only in Whitecliff and Alum bays in the Isle of Wight and Studland Bay on the mainland. The lowest beds record the first marine incursion into the area after the end of the Cretaceous, though the fossils suggest water of less than normal marine salinity. Outside the Woking–Bournemouth line, the lower beds may have been laid down in fresh or brackish water in channels draining sandy flats. The main mass of the formation in that area is probably a terrestrial deposit in which seams of clay containing leaves record the quiescent phases of flash floods. The heavy minerals (Blondeau and Pomerol, 1969) suggest that the sands were derived from westerly sources.

### London Clay Basement Bed

The London Clay Basement Bed rests with a sharp and slightly erosive junction on the Reading Beds. At Castle Hill, near Cranborne, it oversteps directly on to the Chalk. It is generally sandy and glauconitic and in places is packed with the tubes of the marine worm *Ditrupa*. It marks the first marine transgression over the whole area. It was classified as Palaeocene (instead of Eocene) by Stinton (1975) on the basis of its fish otolith fauna, and by Cooper (1976) on other palaeontological grounds.

## Eocene

### London Clay above the Basement Bed

The London Clay above the Basement Bed is about 90 m thick at Whitecliff

Bay, about 70 m at Alum Bay, about 100 m in a well at Christchurch, and about 90 m in the Fareham district. It thins westwards to 27.4 m at Verwood and 36 m at Fordingbridge and is probably less than 20 m thick near Dorchester. Uncertainties arise because of the sandy character of the uppermost beds, in transition to the overlying formation.

The rock is generally a bluish grey clay in the eastern part of the district, with scattered septarian concretions. It becomes siltier and sandier towards the west, and lines of flint pebbles are seen at several horizons. It nevertheless forms wet and heavy land over an outcrop which in places (for example, north and north-east of Southampton) exceeds 1.5 km in width. Where not urbanised it is still for the most part heavily wooded.

The London Clay represents approximately a complete marine cycle of deposition, from shallow, perhaps inter-tidal, conditions at the beginning, to normal shelf-deposition in quiet water, and to inter-tidal conditions again at the end (Murray and Wright, 1974). Complete sections can be studied at Bognor, Whitecliff Bay and Alum Bay, and existed in the past at Portsmouth, Southampton and Fareham. King (1981), in a paper describing the stratigraphy of the London Clay of the London and Hampshire basins, has divided the succession into four transgressive/regressive cycles, usually with pebble beds at their bases. These cycles, which are identified by the letters A, B, C and D, he has correlated from Bognor to Whitecliff Bay, Portsmouth and around the outcrop through Southampton to as far west as Fordingbridge. The correlation is based on the sedimentological sequence of the cycles and also on a variety of fossil groups, including foraminifera, molluscs, nannoplankton, diatoms, ostracods and fish. Another zonal system based on the dinoflagellate *Wetzeliella* has been devised by Costa and Downie (1976).

Detailed observations of the London Clay have been made at Bognor Regis, where Venables (1963) patiently pieced together the following sequence along some 6 km of foreshore:

|  | m |
|---|---|
| Upper Clay | 45.00 |
| Barn Rock Bed | 2.40 |
| Middle Clay (including Aldwick Beds) | 18.30 |
| Bognor Rock group | 9.75 |
| Lower Clay | 26.00 |
| Total | 101.45 |

The most interesting fossil horizons at Bognor are in the Aldwick Beds, namely two fish beds with a beetle bed between them. These yield, apart from fish remains and beetles, a rich flora of seeds and fruits and many planktonic foraminifera.

'Lower Swanwick' is the name given to a succession of brickpits worked for over 70 years to the east of Bursledon. It was here that the first fossil *Rhabdopleura* was found. This genus belongs (with *Balanoglossus*) to the Hemichordates, a group thought to link the extinct Graptolites with vertebrates. The successions observed in these pits over the decades are difficult to relate to each other (James, 1974), but C.A. Wright (1972a) showed that, although most of the beds were laid down under deltaic conditions, one layer—a sandy clay with *Turritella*—yielded abundant planktonic foraminifera. This layer, now termed the 'planktonic datum' (Wright, 1972b), is correlated with the similar horizon at Bognor and with a level 65.5 m above the base of the formation at Whitecliff Bay and 21 m above the base at Alum Bay. It may perhaps coincide with the deposition of fossiliferous London Clay at Clarendon and Verwood and with the overstep of the London Clay near Cranborne—that is, with the maximum extent of the London Clay sea.

The commonest macrofossils in the London Clay are molluscs: gastropods such as *Turritella* and *Ficus*; bivalves such as *Panopea* (Plate 15), *Pholadomya, Pinna, Pitar*, and many more. Crabs (*Palaeocorystes, Zanthopsis*) are found and serpulid tubes (*Rotularia bognoriensis*) may form knotted masses.

Chandler (1964), summing up a life's work on the plant macrofossils of the London Clay, concluded that the conditions on land were like those of present-day tropical rain forests. Gruas-Cavagnetto (1976) believed that the flora was essentially a subtropical one with a plentiful admixture of tropical forms. She envisaged a latitude of about 35° N, which would correspond to a more or less subtropical climate, towards the warm temperate zone, and drew an analogy with present-day conditions in south-east Asia. An alternative explanation was offered by Daley (1972), who pointed out that, given ample moisture and warm equable temperatures, rain forest may extend beyond the latitudes of mean annual temperature of 26° C which it most favours, perhaps even as far as 40° N, which was the probable latitude of Britain in Eocene times. He envisaged river valleys lined by gallery forests like those of present-day Brazil, with higher and drier interfluves supporting a more temperate flora.

Blondeau and Pomerol (1969) suggested a northern derivation for the heavy minerals of the London Clay and interpreted the abundance of smectite among the clay minerals as evidence of the erosion of vast volumes of Chalk; this agrees with other evidence for a low-lying hinterland. It may be doubted whether the volume of Chalk was sufficient, and volcanic sources in the North Sea (Jacqué and Thouvenin, 1975) may also have contributed.

**Bagshot Beds of the Isle of Wight**

Sandy beds between the London Clay and the Bracklesham Beds in the Isle of Wight are known as Bagshot Beds, although it appears from indirect evidence (Curry, 1958, 1965) that they must be older than the Lower Bagshot Beds of the London Basin. They are 42 m thick at Whitecliff Bay and 6.1 m at Alum Bay (Eaton, 1976; see Figure 25), and on the mainland they extend in a narrow belt from Portsmouth towards Fordingbridge. They are yellow and white micaceous sands with seams of grey pipeclay and impersistent bands of flint pebbles.

**Plate 15** Fossils from the Eocene

1 *Panopea intermedia*. 2 *Nummulites variolarius*, x 14, *a* external view, *b* section to show internal structure. 3 *Nummulites laevigatus*, x 7.5, *a* external view, *b* section to show internal structure. 4 *Glycymeris spissa*, *a* external view, *b* hinge and ligament area. 5 *Calyptraea aperta*. 6 *Chama squamosa*. 7 *Crassatella sulcata*. 8 *Lentidium tawneyi*, x 14, *a* external view, *b* internal view. 9 *Olivella branderi*. 10 *Zanthopsis leachi*, *a* dorsal, *b* anterior view. 11 *Venericor planicosta*, *a* right valve, *b* hinge of same.

# Tertiary (Palaeogene) 97

**Table 11** Classification of the Palaeogene Beds of the Hampshire Basin. (The Stage and Substage names are not standardised.)

| Nanno-plankton zones | Planktonic foraminifera zones | Period | Sub-period | Stage* | Sub-stage* | Group | Formation | Member | Bed | Probable time equivalent units |
|---|---|---|---|---|---|---|---|---|---|---|
| 22 | 19 | Oligocene | Lower Oligocene | Rupelian | | | Hamstead | Upper Hamstead Beds or Bouldnor Member | | |
| | 18 | | | | | | | Lower Hamstead Beds or Porchfield Member | | |
| 21 | | Eocene | Upper Eocene | Bartonian | Lattorfian | | Bembridge | Bembridge Marls or Whitecliff Member | | |
| | | | | | | | | Bembridge Limestone | | |
| 20 | 17 | | | | | | Solent | Osborne Member | | |
| | 16 | | | | | | | Middle and Upper Headon Beds or Headon Member | | |
| 19 | | | | | | | | | | |
| 18 | 15 | | | | Ludian | | Barton | Lower Headon Beds or Hordle Member | | |
| 17 | | | | | | | | Upper Barton Beds or Becton Member | G–L† | Barton Sands |
| | 14 | | | | Marin-esian | | | Middle Barton Beds or Naish Member | C–F† | Barton Clay |
| 16 | 13 | | | | | | | Lower Barton Beds or Highcliff Member | A1–3 B† | Hengistbury Beds (part) |

Tertiary (Palaeogene) 99

| Nanno-plankton zones | Planktonic foraminifera zones | Period | Sub-period | Stage* | Sub-stage* | Group | Formation | Member | Bed | Probable time equivalent units |
|---|---|---|---|---|---|---|---|---|---|---|
| 15 | 12 | Eocene | Middle Eocene | Auversian | | Brackles-ham | Selsey | Huntingbridge | XIX‡ | Hengistbury Beds (part) |
| | 11 | | | | | | | | IX–XVIII‡ | Boscombe Beds and Bournemouth Marine Beds ?Creechbarrow Limestone |
| 14 | 10 | | | Lutetian | | | Earnley | | VI–VIII‡ | ?Creechbarrow Beds below limestone ?Bournemouth Freshwater Beds |
| 13 | 9 | | | | | | | | | |
| | 8 | | Lower Eocene | Cuisian | | | Wittering | | I–V‡ | ?Agglestone Grits Pipeclay Beds and Redend Sandstone |
| 12 | 7 | | | Ypresian | | Upper London Tertiary | London Clay | Bagshot Beds of Isle of Wight London Clay above Basement Bed | | |
| 11 | 6b | | | | | | | | | |
| 10 | 6a | Palaeocene | | | | | | London Clay Basement Bed | | |
| 9 | 5 | | | Sparnacian | | Lower London Tertiary | Reading Beds | | | |

* The Stage and Substage names are not standardised.
† Burton, 1933.
‡ Fisher, 1862: type section at Whitecliff Bay.

100   The Hampshire Basin

In Table 11 the Redend Sandstone, Pipeclay Beds and Agglestone Grits of the country around Wareham, which Arkell (1947) placed in the Bagshot Beds, have been classified with the Bracklesham Group: there is no evidence for or against grouping them with the Bagshot Beds of the Isle of Wight except that, on general geometrical grounds, they must at least partly correspond with the Bournemouth Freshwater Beds and thus with the Bracklesham Group. Costa and others (1976) have pioneered a palynological examination of the problems of correlating these beds.

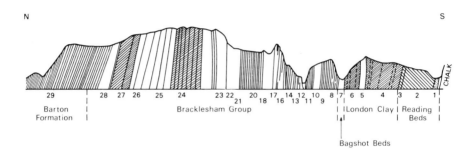

**Figure 25**   Section of Eocene beds in Alum Bay, Isle of Wight
(Gardner, J.S., after Prestwich, 1846, pl. 9, fig. 1, with Prestwich's bed numbers)

**Bracklesham Group**
The stratigraphy of the Bracklesham Group was first described in detail by Fisher (1862), and his work has never been superseded. He chose the cliff and foreshore section in Whitecliff Bay as the type section and divided the group into 19 beds, I to XIX. In the foreshore exposures of Bracklesham Bay and Selsey Bill (Figure 26) he saw neither a complete nor a continuous section. Curry and others (1977) were the first to give reliable measurements of thickness here. Fisher recognised 22 beds (1 to 22), but the lowest of these is equivalent to bed VI at Whitecliff Bay. The higher beds of this section can be correlated with exposures in the New Forest.

Stinton (1975), followed by Curry and others (1977), named the Wittering, Earnley, Selsey and Huntingbridge Formations within the Bracklesham Group; the first two correspond to Fisher's groups D and C. Cooper (1976) reduced the Huntingbridge Formation to a member of the Selsey Formation. The Wittering, Earnley and Selsey formations of his account correspond to the Cuisian, Lutetian and Auversian stages, which are palaeontologically defined. It has not yet been shown that they meet the main criterion for lithostratigraphic formations, namely mappability.

Although the term 'Bracklesham Group' must be defined by reference to the type section, where both marine and continental beds are found, it is here extended to include the more marine sequence east of Southampton Water and the increasingly continental sequences (formerly termed 'Bagshot Beds') at Alum Bay and west of Christchurch.

Tertiary (Palaeogene) 101

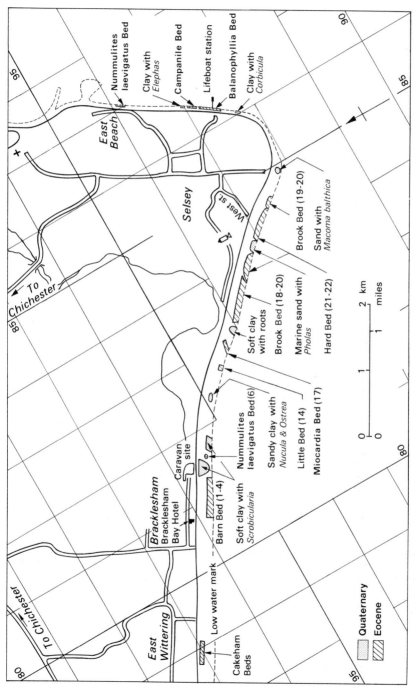

**Figure 26** Foreshore exposures of Bracklesham Group and Quaternary deposits between Bracklesham and Selsey Bill, Sussex
(Curry *in* Curry and Wisden, 1958, fig. 2)

The fauna of the marine Bracklesham Group is very rich and diverse. About 500 species of molluscs are known, mostly indicative of shallow clear warm water; the cool water indicators of the London Clay are absent. Corals, polyzoa, crabs and sea-urchins also occur. Foraminifera are not common, except for species of *Nummulites* (Plate 15) which abound at certain horizons and which provide the key to the stratigraphy of the beds, as shown in Figure 27. Zonal schemes have been constructed recently on the basis of the foraminifera (Murray and Wright, 1974) and the dinoflagellates (Eaton, 1976).

The Bracklesham Group is about 200 m thick at Whitecliff Bay, where the base consists of a bed, 25 to 45 cm thick, of rounded flint pebbles. A composite section at Gosport and Fawley gives a thickness of about 140 m, and here too there is a pebble bed at the base. The incomplete foreshore section at Bracklesham and Selsey measures about 114 m.

The beds and banks of the New Forest streams ('gutters') around Bramshaw are classic fossil localities in the upper part (Huntingbridge Member). In places the banks have been quarried back many metres by fossil collectors. Sections are given in Stinton (1970).

The relationships of the marine and freshwater facies of the Bracklesham Group are sketched in Figure 27. The deposits become increasingly non-marine in a westerly direction, and even at Whitecliff Bay (beds V and XIII) there is a substantial lignitic coal seam and some pipeclay. At Alum Bay, marine shells are found only in the top 11 m or so; at lower horizons, marine influence is attested by glauconite and marine dinoflagellates. There are substantial seams of pipeclay here and again in the Bournemouth Freshwater Beds, where they have been worked in the past. The pipeclay beds of the Wareham district form one of the most important mineral resources of Dorset. Above these lie coarse sands (Agglestone Grits) and gravels composed of well-rounded pebbles—mostly flint, but with some types from Cornubia and Brittany, and also silicified Purbeck limestones. These sands and gravels extend westward as outliers as far as the area around the Hardy Monument, near Abbotsbury, where they reach an elevation of 241 m. West of Wareham the beds are covered by Pleistocene gravels which are distinguished from those of the Eocene mainly by the angularity of the flints they contain.

The Creechbarrow Beds, described in detail by Arkell (1947), are usually correlated with the Bembridge Formation on the basis of the land and freshwater molluscs in the limestone which forms their highest unit. However, borings for pipeclay show that the bulk of the unit is indistinguishable from the Agglestone Grits of the neighbouring heaths, and there is no sign of any sedimentary break at the base of the limestone. The fossils in question are, moreover, long-ranging forms of little reliability in correlation. The beds are accordingly here regarded, following Cooper (1976), as probably having been formed at the same time as the upper part of the Bracklesham Group.

The area of outcrop of the remaining Tertiary formations is much reduced; none is found south-west of a line from Alum Bay through Christchurch to Ringwood, or north of the latitude of Bramshaw.

# Tertiary (Palaeogene)

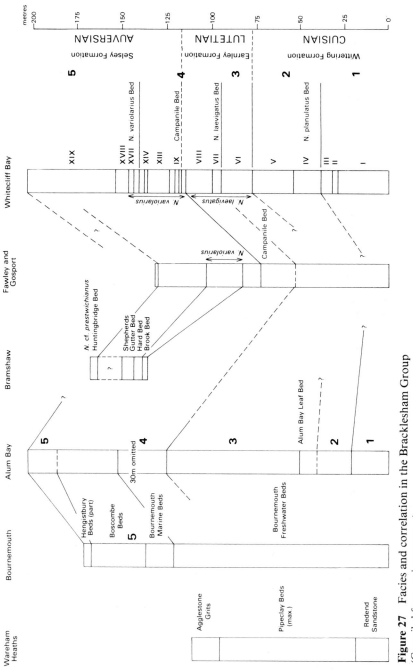

**Figure 27** Facies and correlation in the Bracklesham Group
(Compiled from various sources)
1–5 Microplankton zones of Eaton, 1976
I–XIX Beds of Fisher, 1862

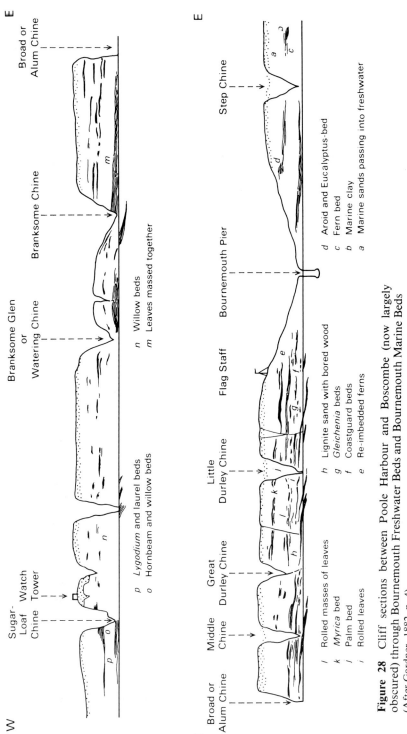

**Figure 28** Cliff sections between Poole Harbour and Boscombe (now largely obscured) through Bournemouth Freshwater Beds and Bournemouth Marine Beds (After Gardner, 1882, p. 4)

## Barton Formation

The type section of the Barton Formation, (Figure 29) including the beds formerly called Lower Headon Beds, is in the cliffs that run from Cliff End, 1 km E of Mudeford, to Paddy's Gap, near Milford. The formation is here 67.5 m thick; it thickens south-eastwards to 102 m at Alum Bay and 111 m at Whitecliff Bay. Curry (1976) argued that the Hengistbury Beds, of which the position in the sequence is not clear, should be correlated with the lower part of the Barton Beds.

Geological Survey maps show two divisions—Barton Clay overlain by Barton Sands—below the Lower Headon Beds. Burton (1933) identified 14 divisions which can all be recognised only in the type section. They can, however, be grouped into Lower, Middle and Upper divisions which can be identified at Alum Bay. These, with the Lower Headon Beds, are now termed the Highcliff, Naish, Becton and Hordle members.

The general lithic characters and most prominent fossils of the Barton Formation are shown in Table 12. The base of the formation at Cliff End is taken at a pebble bed which separates sands beneath from 3 m of clay above. These in turn are followed by glauconitic sandy clays (Bed A1 of Burton) with *Nummulites prestwichianus*, indicating a more open sea with more turbulent conditions. Bed A2 contains *Nummulites rectus* 6 m above *N. prestwichianus*; this is the last known nummulite in Britain. Some 600 species of molluscs are known from the formation; beds C, E and H are the most fossiliferous. Bed E alone yields some hundreds of species, including *Cornulina minax, Clavilithes macrospira, Sycostoma pyrus, Crassatella sulcata* (Plate 15) and *Cardiocardia sulcata*.

The boundary between the Barton Clay and the Barton Sands corresponds with that between Burton's beds F and G. Yet the fossils indicate that the sea, shallow at the beginning of Barton times, continued to deepen until the time of Bed H, the Chama Bed, though occasional stringers of estuarine shells and pockets of seeds show that land was never far away. Bed I yields no fossils. Bed K shows the first signs of a progressive change through brackish to freshwater conditions in the Hordle Member. This change came about not so much by a fall in sea level as by a silting up of the sea and its replacement by freshwater lagoons and marshes. In the Isle of Wight the top of the Hordle Member at Headon Hill and Colwell Bay is marked by a massive cream-coloured freshwater limestone, the How Ledge Limestone, 0.9 to 1.4 m thick, and largely composed of the freshwater gastropods *Galba* and *Planorbina*.

In Hordle Cliff on the mainland, the Leaf Bed flora suggests a coastal alluvial plain with swamp forests, like present-day Florida, at the eastern end of the outcrop; the flora from the western end includes montane elements related to species found today in southern China. Both pieces of evidence suggest a distinctly cooler climate than obtained during the deposition of the London Clay.

## Solent and Bembridge formations

The Solent and Bembridge formations consist of some 150 m of freshwater deposits (fluviatile, marsh and lagoonal) with thin deposits representing two brief marine incursions.

The Solent Formation comprises the Headon Member (formerly the Mid-

106  *The Hampshire Basin*

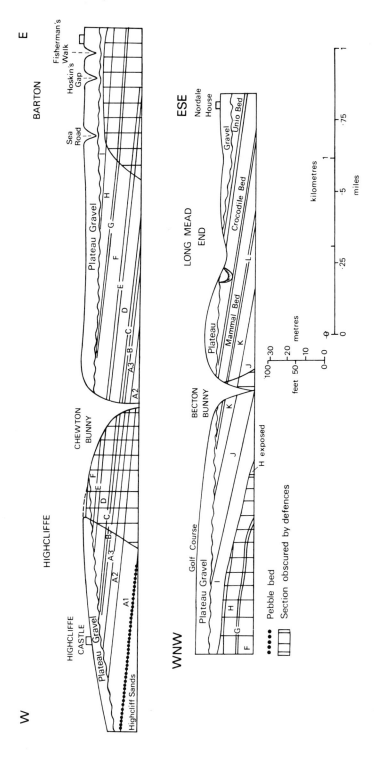

**Figure 29** The Barton Cliff section
Vertical lines indicate coast defences (From Hooker, 1975, p. 167)

# Tertiary (Palaeogene)

**Table 12  Stratigraphy of the Barton Formation**

| Member | Bed | | metres | Description |
|---|---|---|---|---|
| Lower Headon Beds, or Hordle Member | | Rodent Bed | 1.5 | Pale brownish marl. *Lymnaea, Theridomys* |
| | | Unio Beds | 5.2 | Laminated pale and dark grey clays with sandy layers. *Unio solandri, Viviparus lentus*, seeds of water plants |
| | | | 5.5 | Pale blue and green clays with seams of lignite, thin limestone near base |
| | | Chara Beds | 1.2 | Laminated sandy clays with pockets of *Chara* nucules |
| | | Crocodile Beds | 3.7 | Brown sand with white silt at top. *Erodona gregarea, Potamides vagus*, fish scales, remains of crocodiles, turtles and mammals. |
| | | Leaf Bed | 0.9 | Purple sands and laminated clay with pockets of leaves and seeds. |
| | | Mammal Bed and basal clays | 5.2 | Pale green clays, layers of white sand with *Viviparus lentus*. Mammals rare, Ironstone nodules in basal clays. |
| Upper Barton Beds, or Becton Member | L | | 1.2 | Black clays, crushed shells |
| | K | Long Mead End Bed | 6.0 | Pale sands with brackish-water shells at top. *Batillaria concava, Bayania fasciata, Corbicula* |
| | J | Becton Bunny Bed | 8.0 | Grey-brown clays with both marine (*Olivella, Pollia, Nucula, Pitar*) and estuarine shells (*Potamides, Corbicula, Bayania*). *Callianassa* in soft concretions |
| | I | | 8.0 | Sands, unfossiliferous |
| | H | Chama Beds | 5.5 | Sandy clays. *Chama squamosa, Crassatella tenuisulcata, Glycymeris deleta, Cardita oblonga, Conorbis dormitor, Hemiconus scabriculus, Pollia lavata* |
| Middle Barton Beds, or Naish Member | G | Stone Band | 0.3 | Limestone made of comminuted shells. *Turritella, Tellina* |
| | F | | 6.0 | Brownish grey clay. *Turritella, Corbula* |
| | E | Earthy Bed | 1.5 | Glauconitic sandy clays, septaria at top. Large shells (see text) |
| | D | | 6.0 | Glauconitic clays |
| | C | | 1.8 | Glauconitic sandy clay, septaria at top and bottom. *Athleta suspensus, Clavilithes longaevus, Callista belgica* |
| | B | | 1.2 | Grey clays. *Pholadomya* |
| Lower Barton Beds, or Highcliff Member | A3 | Highcliff Sands | 3.0 | Stiff grey clay with beds of grey sand. *Sconsia ambigua, Bartonia canaliculata, Dientomochilus bartonensis* |
| | A2 | | 4.0 | Glauconitic clays. *Turricula lanceolata, Athleta athleta, Nummulites rectus*, otoliths, sharks teeth. |
| | A1 | | 6.0 | Glauconitic sandy clays. *Nummulites prestwichianus* at base |
| | | | 3.0 | Clay |
| | | | | Bed of flint pebbles |

dle and Upper Headon Beds) and the Osborne Member (formerly Osborne Beds). The Brockenhurst Bed and Venus Bed at the base of the Headon Member are marine, and richly fossiliferous. The base rests on an uneven surface of the Hordle Member and in places contains fragments of that member as well as flint pebbles. The fauna includes a dozen species of corals (e.g. of *Solenastraea* and *Lobopsammia*) and many molluscs closely matching those found at Latdorf, Mecklenburg, in Germany. The gastropod *Athleta dunkeri* is prominent among these. This 'Lattorfian' fauna was formerly taken to mark the beginning of Oligocene time, but it is now placed as Upper Eocene. The upper part of the Brockenhurst Bed consists of some 12 m of sandy clays in which alone the gastropod *Neoathleta geminatus* occurs.

The Venus Bed (the name fossil is now called *Pelycora suborbicularis*) consists of 4.5 m of pale blue and greenish clays and pale sands at Headon Hill (10 m at Whitecliff Bay) with fossils showing a gradual retreat of the sea under estuarine conditions (*Potamides, Corbicula, Hydrobia*). At Colwell Bay the bed has been washed out by a channel filled with green clay including an oyster-lumachelle (*Ostrea velata*) like those of the Middle Jurassic Fuller's Earth. This oyster bank is some 3 m thick and 30 m wide and wedges out at either end of its 0.8-km outcrop. The whole of this marine and estuarine part of the Headon Member is 38.4 m thick at Whitecliff Bay, 10 m at Headon Hill, and 5.6 m at Lymington.

The remaining Palaeogene beds are known only in the Isle of Wight, except for patches of upper Headon Member to the south of Lyndhurst.

The upper part of the Headon Member is 17.7 m thick at Whitecliff Bay and 14.2 m at Headon Hill. It consists mostly of green and brown clays with some thin sands, but also includes the Headon Hill Limestone, 4.3 m of yellowish limestone with interbedded lignite and clay, forming a prominent feature between Alum Bay and Totland Bay. Brackish or freshwater shells are common, except in the sands.

The Osborne Member is 24.4 m thick at the east and west ends of the Isle of Wight and 33.5 m in the north. It consists of brackish and freshwater clays and marls, some highly coloured, and discontinuous bands of concretionary limestone. Shells, of the same species as in the Headon Member, are common, especially in some of the limestones, in which nucules of lime-secreting aquatic plants (Charales) are abundant. Fishes are found in some beds, especially *Diplomystus vectensis* in a thin bed of clay exposed on the shore west of Wootton Creek. Mammals (*Palaeotherium, Theridomys*) occur in other beds. On the north-eastern shores of the island the member is represented by a lower unit of coarse, indurated sands (Nettlestone Grits) overlain by the St Helens Sands.

The Bembridge Formation comprises the Bembridge Limestone below and the Bembridge Marls above. The limestone is 5.5 m thick in the west of the Isle of Wight and 8 m thick in the east, but borings show it to be much thinner where it is concealed beneath the Hamstead Formation, particularly near Ryde. The massive beds of pale limestone with intercalated greenish clays are easily recognised at outcrop and are accessible at Whitecliff Bay. They yield the same freshwater gastropods as the Headon Member and the large land shells *'Filholia' elliptica* (Plate 16) and *Megalocochlea pseudoglobosa*. The mammal fauna—with species of *Palaeotherium, Cetochoerus* and

*Anoplotherium*—is the same as that from the gypsum deposits of Montmartre, studied by Cuvier at the beginning of the 19th century. Charalean nucules are common.

The Bembridge Marls are 26 m thick in the east of the Isle of Wight and 21.3 m in the west. At the base is a 3-m bed of sand packed with oysters encrusted by serpulids and barnacles. This represents the second of the two marine incursions mentioned on p. 105, but it does not extend west of Bouldnor cliff and is more estuarine than fully marine in character. The bed contains pebbles of fresh flint, and most of the shells are broken either by water movement among the pebbles or by the crushing teeth of rays (*Myliobatis*). The fauna is too restricted in diversity to demonstrate any particular correlation with marine beds on the Continent. Hancock and Curry (1977) took this bed as the provisional base of the Oligocene in England.

The succeeding marls represent a regime of brackish lagoons and of floodplain lakes and marshes with a freshwater fauna. Seeds and leaves of water plants show that the water seldom exceeded four or five metres in depth. Sedimentary structures indicate that freshets occasionally eroded channels in the lake floors (Daley, 1973).

The marls show cyclic repetitions of conditions of deposition. After an episode of river erosion, laminated muds and silts were laid down. These become finer-grained upwards and the higher layers contain freshwater snails characteristic of slow-flowing water and drifted remains of insects belonging to at least 14 orders. Each cycle ends with colour-mottled muds with rootlets representing fossil soils, some waterlogged, others exposed to the air. These muds were truncated by renewed downcutting by a river. Daley interpreted these cycles as evidence of alternating episodes of high and low rainfall.

The Bembridge Formation shows the first evidence of tectonic activity in the area in Tertiary times (Daley, 1971). The variations in thickness of the Bembridge Limestone and of certain recognisable beds in the Marls are due to warping during deposition. The fresh flints in the Oyster Bed show that exposed Chalk was being eroded nearby.

## Oligocene

### Hamstead Formation

At the type-section in the cliffs between Hamstead and Bouldnor, the Hamstead Formation, which succeeds the Bembridge Marls by a gradual transition, is 77.7 m thick. The beds include coloured clays, loams, sands and shales, mostly of freshwater origin. A third marine phase not mentioned on p. 105 is recorded in the uppermost beds. The transition to marine conditions must have been very gradual, for there is no basement bed, or any trace of the erosion of underlying beds that marks all the other Tertiary marine transgressions. The marine fauna (Plate 16) is, however, sufficient to prove a correlation with marine deposits of Lower Oligocene age on the continent.

The following is a generalised sequence:

*Upper Hamstead Beds* m
  Marine clays (Corbula Beds) with *Corbula subpisum, Pirenella monilifera, Ostrea callifera, Neoathleta rathieri,* etc.  5.3

Upper estuarine and freshwater beds (Cerithium Beds) with *P. monilifera,
Nystia duchasteli, Viviparus lentus, Unio, Corbicula,* etc.                    3.7

*Lower Hamstead Beds*
Middle estuarine and freshwater beds, with *Bayania fasciata, Viviparus,
Planbris, Unio, Nystia,* etc., and carbonaceous seams with plant
remains. White Band (shelly clays) at base                                      48.4
Lower estuarine and freshwater beds, with *Melanoides acutus, Stenothyra
pupa, Galba, Planorbis, Sphenia minor, Polymesoda convexa,* etc.
Black Band, full of *Viviparus lentus* and *Unio,* at base                     20

Ostracods and plant-remains also occur. The mammals (*Bothriodon* and *Entelodon*) are those of the *Calcaire de Brie* in the Paris Basin.

Here ends the record of Tertiary times in the area. If later Oligocene, or still younger Tertiary formations were laid down, no trace remains. The next stage in the history of the region is one of uplift, tectonic movements, and denudation. The next deposits to be seen are of Quaternary age.

**Plate 16** Fossils from the Eocene and Oligocene
1 *Crassostrea vectensis.* 2 *Sycostoma pyrus.* 3 *Clavilithes longaevus.* 4 *'Filholia' elliptica.* 5 *Pelycora suborbicularis, a* left valve, *b* anterior view. 6 *Typhis pungens.* 7 *Glycymeris deleta, a* external view, *b* hinge and ligament area. 8 *Corbula pisum,* x 14. 9 *Galba longiscata.* 10 *Stenothyra parvula,* x 14. 11 *Viviparus lentus.* 12 *Athleta (Volutospina) luctator.* 13 *Neoathleta rathieri.* 14 *Potamides mutabilis.* 15 *Batillaria concava.* 16 *Nucula headonensis.* 17 *Calliostoma nodulosum.* 18 *Potamomya plana, a* external view, *b* internal view. 19 *Potamaclis turritissima.* 20 *Rimella rimosa, a* dorsal view, *b* oral view. 21 *Fusinus porrectus.* 22 *Corbicula obovata, a* left valve, *b* dorsal view.

Tertiary (Palaeogene)

# 6 Structure

## Introduction

Until comparatively recently, the Hampshire Basin with the adjoining areas of Dorset and Wiltshire was thought of as a practically undivided basin of continuous subsidence throughout Mesozoic and early Tertiary times. However the evidence from boreholes sunk in search of hydrocarbons during the last few decades and that derived from geophysics shows that this picture must be extensively revised. The floor of the major basin of deposition (Figure 30)—of which the shape is well picked out by the outcrop of the base of the Tertiary rocks—is divided into a number of concealed subsidiary basins, or grabens, separated by blocks of stable material, or horsts (Figure 31). The outlines of these ancient features can be detected in the pre-Mesozoic rocks, and their presence affected sedimentation in Mesozoic and later times and also controlled the position and alignment of folds and faults visible at the surface.

Geophysical investigations have used two main methods. The first examines variations ('Bouguer anomalies') in the strength of the Earth's gravity above or below a norm. The strength is less than normal ('negative anomaly') over thick accumulations of lighter rocks (as in the grabens and over acid igneous intrusions) and nearer to or above normal ('positive anomaly') where heavier rocks are nearer the surface (as over horsts and basic igneous intrusions). In this area, all the anomalies are negative to a greater or lesser extent. Measurements of gravity are supplemented by seismic methods, which record the speeds at which sound travels through the rocks. Deep boreholes enable the rocks whose physical properties have been measured in these ways to be identified in geological terms.

The scanty evidence of deep geology within the region shows that the best indications of the likely deeper structures are to be found in south-west England. There, Devonian and Carboniferous sediments were tightly folded by intense northward compression in the Variscan orogeny towards the end of Palaeozoic times. Since then they have behaved as a relatively rigid mass in which folding has not been reactivated, but faulting, mostly of a tensional character, was the dominant structural factor in Mesozoic times. Northwards compression and folding, with reversed faulting, was resumed in the Alpine orogeny in Miocene times, but it did not affect the deep-seated rocks or their structures.

Most of the area is underlain by Devonian and Carboniferous rocks similar in facies and in structural pattern to those of south-west England. Only in the northern and extreme eastern parts are there rocks present in the shelf-facies that they show elsewhere in England and Wales. The boundary between the two provinces, may be a very ancient fault-line, running south of and parallel to the Hercynian front. South of this, in the Variscan province, the structures can be divided into two main groups, those trending approximately east–west, and those with a roughly north-west–south-east course.

Structure 113

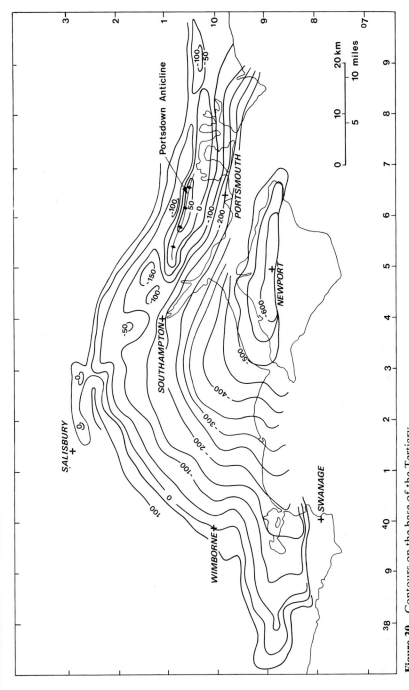

**Figure 30** Contours on the base of the Tertiary
Heights in metres relative to Ordnance Datum (by Mrs R. Moseley)

## Structures trending approximately east–west

Faulting of east–west trend was probably initiated late in the Variscan orogeny and its effects are widespread in Devonian, Carboniferous and New Red Sandstone rocks in south-west England. Whittaker (1975) postulated rift-valley systems bounded by east–west faults in southern England, and projected one of them, the Central Somerset graben, across the northern part of the area, the present surface expression of the southern margin of the graben being the Mere Fault (Figure 31). Another rift, the Crediton Trough, can be tentatively traced through an east–west zone of thick New Red Sandstone and Lias beneath the Marshwood Dome and eastwards towards Wimborne Minster. Other faults such as the Bridport and Beaminster faults and the Lopen–Chiselborough–Coker fault system south of Yeovil may also have deep-seated origins.

In the horsts between the grabens the pre-Mesozoic strata come relatively close to the surface. One such structure, the Fordingbridge 'high', was first recognised by gravity survey (White, 1948); in it Devonian slates are known to be within 1600 m of the surface in the Cranborne Borehole. White's work showed a pattern of roughly east–west positive and negative anomalies which may correspond with fault-bounded blocks at depth. Their surface expression may be seen in the many east–west folds that occur in our region. These vary from gently asymmetric anticlines, such as the Winchester Anticline, to more sharply monoclinal folds such as the Isle of Wight Monocline and its westerly extension, the Purbeck Monocline (Figure 31).

Until recently these asymmetric folds were assumed to be due to northwards compression during the Alpine orogeny, but it is now thought that they may have resulted simply from the draping of plastic Mesozoic and Tertiary strata over active basement faults, and hence to be related to tensional movements. Nearly all the folds in the Hampshire Basin face northwards, indicating that the latest movement on the concealed faults was a downthrow to the north. However, one of these folds, the Isle of Wight Monocline, finds its mirror image in a south-facing east–west monocline in the English Channel about 60 km to the south (Figure 31). This suggests that the area between these outwards-facing folds formed a horst, at least as far as the latest of the Tertiary movements is concerned.

The history of the movements of the east–west faults is complex, involving in some cases reversal of direction. This applies particularly to the structures in the Isle of Purbeck, which have attracted considerable controversy. These comprise such elements as the Weymouth and Purbeck anticlines, the Sutton Poyntz, Poxwell and Chaldon periclines, and the Ridgeway and Abbotsbury faults (Figure 33). Arkell (1936, 1947) and Taitt and Kent (1939) suggested a possible reversal of movement on the Abbotsbury (Poxwell) fault, the earliest movement having been a downthrow to the south. Arkell (1947) and House (1961) interpreted the faulting associated with the steep limbs of the Poxwell and Chaldon periclines as showing a reversed movement in Tertiary times which counteracted the earlier, normal pre-Albian movement at depth (Figure 32b). More recently Ridd (1973) postulated that the principal younger reversed faults started as nearly horizontal planes in the plastic Kimmeridge Clay and developed northwards to pass into and shear the older faults which had already been bent over to the north by compressional forces (Figure 32a).

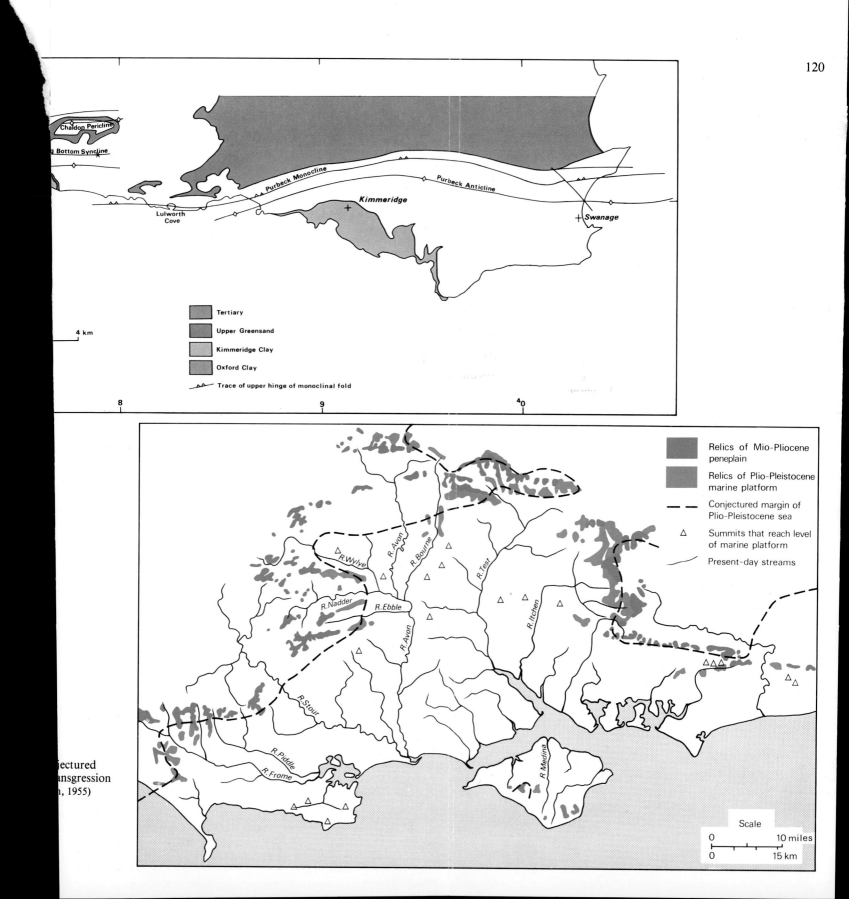

# Structure 115  116 The Hampshire Basin

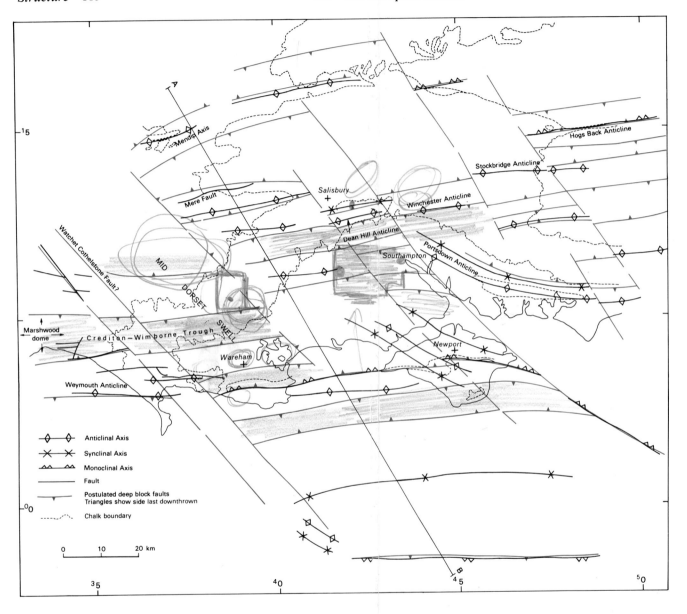

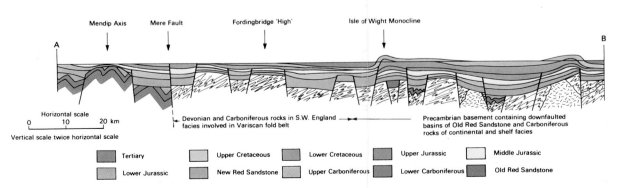

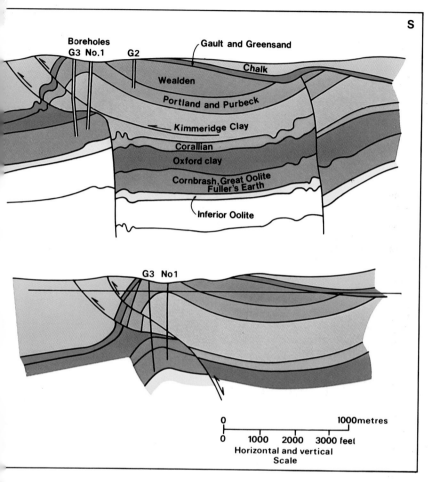

2 Interpretation of the structures associated with the Chaldon Pericline
idd, 1973; b after House, 1961)

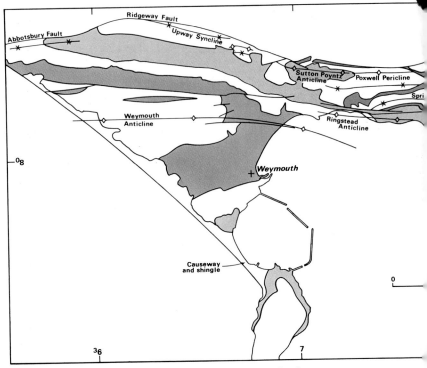

**Figure 33** Major structural elements of the Weymouth–Swanage area (Partly after House, 1961)

**Figure 34** Relationship of present drainage to Plio-Pleistocene marine (After Wooldridge and

Whichever interpretation is nearer the truth, there is clear evidence that the Tertiary compression caused relatively low-angle reverse faults with overriding to the north. Farther east the compression was less intense, and the steep monocline of the Isle of Wight appears to be a drape over a fault throwing down to the north.

One of the most striking features associated with the Isle of Purbeck structures is the 'Lulworth Crumple'. This is well seen in the Purbeck Beds on both sides of Lulworth Cove and at Stair Hole (Plate 9.1). Arkell (1947) interpreted this as a drag fold, with movements upwards and southwards against the dip towards the crest of the main Purbeck anticline. This is inevitable when such thinly bedded and incompetent strata as these are constrained between such massive units as the Portland Beds and the Chalk in a steep fold. West (1964) explained the location of the crumples here and at Peveril Point as due to the difference between the shearing forces at the base and top of incompetent beds involved in disharmonic folding. In this process, there is a sliding movement along the bedding planes between the competent and incompetent units. At the base, the shearing forces reinforce each other, whereas at the top they cancel out.

Some features of the Lulworth Crumple suggest gravity folding. This could have developed later, in a tensional phase, in which the Mesozoic rocks sagged in a drape over a normal fault in the basement rocks.

## Structures trending approximately north-west–south-east

Faults with a north-west–south-east trend are well known in the south-west of England and include the Sticklepath Fault Zone and the Combe Martin Fault. The most easterly known fault of this group, the Cothelstone Fault, was described by Whittaker (1972) who related it tentatively to the Hooke and Winterborne faults north-east and east of Bridport. A steep gravity gradient trending north-west–south-east and sloping north-east is present to the north-east of these faults and probably indicates a deep-seated structural line cutting basement rocks and passing out to sea in the vicinity of Lulworth. The disposition of the Poxwell Fault and its associated folds *en échelon* with the Isle of Purbeck Monocline may be due to a dextral wrench movement in the basement along a line of extension of the Cothelstone Fault.

The Bouguer anomaly map shows many of the east–west features truncated and displaced along north-west–south-east lines. The Fordingbridge 'high', for example, appears to be truncated at its western end, and the gravity contours here and to the south have a strong preferential trend in this direction. A major north-west–south-east structural line crosses the English Channel and is known as the Bembridge–St Valéry line. Curry (1976) believed that this structure swings round to join the Isle of Wight Monocline and suggested that wrench movement on the former was transformed into faulting below the monocline. There is, however, little evidence of lateral movement along the east–west structure. There are nevertheless small asymmetric north-west–south-east folds under Southampton Water, and the major east–west structure of the Portsdown Anticline swings into a north-west–south-east direction under Southampton and continues along the line of the Avon valley, cutting off the Dean Hill anticline-syncline pair. This line probably belongs to a suite of faults that originated in the Variscan orogeny

## 122  The Hampshire Basin

and that affected the pre-Mesozoic rocks, the New Red Sandstone and some Jurassic rocks in southern England, the southern North Sea Basin, and Brittany.

Daley and Edwards (1971) suggested that small north-west–south-east folds in the northern part of the Isle of Wight and on the adjacent mainland were of late Palaeogene date, that is, before the Alpine orogeny and the development of the Isle of Wight Monocline. Other small folds with this trend have been detected offshore about 40 km SSE of the Isle of Purbeck, where they affect Jurassic, Cretaceous and Tertiary beds.

### Influence of structure on sedimentation

Some variations in thickness and facies of beds may be due to the slow, gentle epeirogenic movements described in Chapter 2. In other cases, however, there is no doubt that contemporaneous movement along deep-seated fault lines had a controlling influence on these features. Hallam and Sellwood (1976) demonstrated strong control of sedimentation in the North Sea region by tensional faulting of various trends, and exploration for hydrocarbons in the English Channel may reveal a similar pattern there.

The influence of the east–west faults in the area can be recognised by tracing an identifiable package of rocks across the lines of the grabens and horsts in the basement. Thus the thickness of the Jurassic rocks between the base of the Inferior Oolite and the base of the Oxford Clay varies from 434 m in the Lulworth Banks Borehole, to around 200 m N of the Purbeck fold at the Wytch Farm boreholes. The package thickens again over the postulated New Red Sandstone trough near Wimborne Minster to 292 m in the Winterborne Kingston Borehole and then diminishes again over the Fordingbridge 'high' to around 170 m in the Fordingbridge and Cranborne boreholes.

Data are more sparse in the eastern part of the Hampshire Basin, but the same package of strata is 204 m thick in the Kingsclere Borehole, which is situated over a gravity 'high' and is probably near the fault-bounded edge of a horst which continues eastwards towards Guildford. Boreholes on the Winchester Anticline show 394 m, suggesting that the anticline is situated over a graben with its northern boundary fault underlying the steep northern limb of the fold. The western end of this concealed structure may be defined by the closure of the negative anomaly lines 3 or 4 km S of the axis of the Dean Hill Anticline. The thickness of 241 m in the Portsdown Borehole suggests another graben lying under and to the south of the Portsdown Anticline, and this may be associated with a negative gravity anomaly passing under the estuary of the Hamble River in an east–west direction.

It was said in Chapter 4 that the Wealden sediments were probably derived from an ancient massif extending from Spain to Cornubia (Allen, 1972). This massif may well have included fault-bounded horsts in the English Channel. In the Eocene, the red beds of the London Clay and Bagshot Beds of Alum Bay and Studland give evidence of a nearby area of strong subaerial erosion in more oxidising conditions than those that prevailed in the low-lying marshy shores and shallow seas to the north and north-west. This erosion may have been taking place on the horst already envisaged (p. 114) immediately to the south of the Isle of Wight–Purbeck monocline.

## Structural history of the Hampshire Basin

Towards the close of the Variscan orogeny, compression gave way to tension, and east–west grabens and horsts developed south of the Hercynian front (which may have been a much more ancient feature). The grabens, which were fault-bounded, confined the areas of thick New Red Sandstone deposition to east–west belts which were cut off, and in some cases laterally displaced by north-west – south-east faults (some of which may also follow very ancient lines). Distension in this fault pattern—possibly associated with the crustal tension caused by the opening of the north Atlantic—continued in the Jurassic and Cretaceous with some faults breaking through the Mesozoic rocks and others merely causing the rocks to drape over them.

The pre-Albian transgression (p. 74) indicates a burst of tectonic activity in the area and may correlate with similar movement in the North Sea region. The direction of overstep in the central Hampshire Basin was approximately north-westerly and may be connected with rotational tilting along the north-west–south-east faults. Drummond (1970) produced strong evidence for control of Albian and Cenomanian sedimentation by buried north-west–south-east faults, including the Cothelstone Fault, elements of which he projected into the Lulworth area. The mid-Dorset Swell may be bounded on its north-eastern side by the fault or faults which appear to truncate the Fordingbridge 'high'.

There is considerable evidence of compression during the Tertiary. The reactivation of the Variscan north-west–south-east faults as dextral wrench faults, which can be demonstrated at the surface in the case of the Sticklepath Fault in Devon, suggests compression rather than block movements under tension. Compression is also indicated by thrusting and reverse faulting in the Purbeck area. This phase of tectonic activity, which probably had its origins in the Alpine orogeny, was probably responsible for the general bowl-shaped warp of the Hampshire Basin and for other gentle east–west warps. With the final relaxation of pressure further movements took place on both the east–west and north-west–south-east faults, producing draping of the Mesozoic and Tertiary rocks over them.

# 7 Pleistocene and Recent

## Introduction

After the deposition of the Hamstead Beds the Alpine mountain-building episode in mid-Miocene times completed the existing structural pattern of the region. The history of that time, and of the succeeding Pliocene and Pleistocene periods, is largely a matter of conjecture. The only material evidence is provided by the Superficial or Drift deposits, which are of variable but usually small thickness and irregular discontinuous distribution. In consequence they provide only patchy and disconnected glimpses of the sequence of events. It is appropriate, therefore, to describe the principal kinds of Drift deposits before discussing the history of the period in which they and the existing drainage pattern were formed.

## Clay-with-flints

Scattered over the Chalk outcrop are accumulations of reddish brown or chocolate-coloured clay in which are unworn flints, rounded flint pebbles, and a variable proportion of sand. This is the Clay-with-flints, which ranges in thickness up to about 10 m. It is found both on the surface of the Chalk and filling solution pipes and hollows in it at levels ranging from 90 to 275 m above Ordnance Datum.

In solution pipes and hollows the Chalk surface is commonly lined with a layer of unworn flints stained black or green on the outside, embedded in a few centimetres of black unctuous clay. This is taken to be a residue of dissolved Chalk. More generally, however, the Clay-with-flints includes much material derived from the Reading Beds, and its presence at levels as much as 150 m below the former position of the base of this formation points to the continuation of solution after the removal of the Tertiary strata.

No material foreign to the district has yet been recorded *in situ*, and there is thus no direct support for the suggestion (Kellaway and others, 1975) that the Clay-with-flints is a decalcified till.

On the higher parts of the Chalk downs of the Isle of Wight and south Dorset is a structureless gravel composed of unworn and broken flints, either with no matrix or set in a matrix of quartz sand, flint powder, or chalk rubble. Rounded flint pebbles are occasionally seen and the matrix passes in places into a reddish brown clay. This angular flint gravel may be regarded as a variant of the Clay-with-flints formed in areas where the Tertiary cover was absent or was never thick.

## Plateau Gravels, Valley Gravels, Alluvium

The Plateau Gravels and Valley Gravels are composed almost exclusively of flint with some pebbles of sarsen (silicified sandstone); pre-Cretaceous pebbles are present in small amounts, but there is no proof that these have been transported from outside the region. Plateau Gravels cover a wide range of gravel spreads; some of the highest appear to mark the faint outlines of an early, eastward-flowing river, the 'Solent River' of Reid (1902), while the lowest are clearly related to the existing drainage pattern.

Valley Gravels largely consist of the resorted constituents of the Plateau Gravels. They form terraces lining the valley sides up to 15 m above the present valley floors. In addition they underlie the alluvium of the major streams which, in their lower reaches, are underlain by buried valleys at least 18 m deep.

Alluvium is the modern deposit of the rivers. It is generally a grey silt with seams of gravel or of freshwater shells. Peat is locally associated with it, and may pass laterally into it, particularly where the alluvium occupies low-lying marshy ground. The alluvium passes laterally into shelly estuarine muds and silts in the tidal reaches of the rivers.

## Brickearth, Coombe Rock, Head

Brickearth is a largely unstratified mixture of fine-grained quartz sand or flint sand with clay; finely divided chalk and scattered small flints may be present. Its composition is commonly determined by the formation on which it rests. Several mechanisms have contributed to the deposits: some are solifluction flows on hill slopes; some are flood deposits on flood plains of rivers; some are wind-blown loess; and many have formed by a combination of these processes. A cold climate is necessary for the formation of loess and for the operation of solifluction, but is not indispensable to the formation of all brickearths.

Coombe Rock is a structureless accumulation of chalk rubble and flints in a chalky paste, which may be cemented or not. It is mostly found on Chalk slopes as a mantle up to 5 m thick, and in dry-valley bottoms, where it may be closely associated with brickearth. In places two or more phases of deposition can be recognised.

In some chalk exposures, lines of flint nodules can be traced through the weathering profile and downslope into Coombe Rock. The deposits were formed by the freezing and thawing of saturated chalk. In some localities they seem to have remained in place, but in most they have clearly been moved by solifluction and can be considered as a particular facies of Head.

Head is defined as weathered and broken-up material that has moved downhill by solifluction, though it may pass without interruption into unmoved material. It occurs alike on plateaux, hill slopes and valley bottoms; it may pass laterally into, or be interbedded with, terrace gravels and alluvium. Downwash deposits, which are still in process of formation, are also widespread.

## Raised beaches and submerged forests

Raised beaches, marking former high stands of the sea during Quaternary interglacials, have been recognised at several levels and are broadly correlatable with the terraces in the river valleys. Deposits of shallow marine sand resting on a chalk platform backed by a cliff-like feature at about 30 m OD, constitute the so-called '100 ft' raised beach of West Sussex. It has been suggested that this '100 ft' beach involves two beaches at closely similar levels. Traces of other beaches at about 15 m and 5 m are preserved at Portland Bill and Selsey Bill respectively.

Submerged forests are found in and offshore from the buried mouths of rivers and mark pauses in the rise of sea level during Flandrian (Recent) times.

## Chesil Bank and The Fleet

The Chesil Bank 'is claimed as one of the wonders of the world' (Macfadyen, 1970). It is a shingle bank (Plate 17.1) from 36.5 to 259 m wide, 7.6 to 12 m high, and 29 km long, extending from Burton Bradstock to the Isle of Portland. For the north-westernmost 13 km it forms the beach; for the next 13 km it is a bar separated from the land by The Fleet, a shallow lagoon 200 to 1000 m wide; and in the south it links the mainland to the Isle of Portland. At any one point on the bank the size of the pebbles is remarkably uniform, but along its length the pebbles vary considerably in size; at the north-western end they are the size of grains of maize, at Abbotsbury the size of broad-beans and at Fortuneswell the size of swans' eggs. Offshore, however, the pebbles are graded in the reverse direction. They are nearly all of flint, with some Upper Greensand chert and local Jurassic rocks; small amounts of Triassic and older rocks derived from the west are also present.

Arkell (1947) thought that the bank originated as a bay bar that had extended across what is now West Bay to join an extended Portland Bill to some lost headland south of Beer Head; advancing seas of the Flandrian transgression pushed the bar northwards and flooded the low-lying land, of which the surviving remnant underlies The Fleet. He thought that no new material had been supplied to the bank since that time. It may be significant that Macfadyen (1970) recorded a marked diminution in the average size and weight of pebbles compared with samples taken from the same localities in 1898.

The Fleet is 0.3 to 1.2 m deep. It is fed with fresh water by streams from the mainland, and with salt water by percolation through the shingle. Baden-Powell (1930) adopted an earlier view that it represented an ancient river that had drained into Weymouth Bay before the advance of the Chesil Bank, but the distribution of gravel on the flanking slopes hardly supports this.

## Chronology

Dating the variety of Drift deposits in the region is difficult, the more so because it lay beyond the conventionally accepted limits of the ice-sheets that covered much of Britain on several occasions during the Pleistocene and so has none of the tills that provide useful datum-planes elsewhere. Instead, the variations in climate and changes in sea level that occurred as each successive glaciation waxed and waned provided many repetitions of the conditions under which each particular lithology accumulated. As yet, fossils are of little help in unravelling the sequence, for records are too sparse to warrant comparisons with well-calibrated sequences elsewhere. In these circumstances interpretation is necessarily speculative and makes use of geomorphological evidence whose correct interpretation is open to dispute.

Earth movements were in progress during the Miocene, and it has been suggested that by Pliocene times the region had been worn down to a peneplane on which the major geological structures had no physical expression. The remnants of this peneplane now rise locally to heights of around 300 m, and

1  Chesil Beach, from West Cliff, Isle of Portland. Weymouth Bay is on the right of the picture, separated by a causeway from The Fleet. (A.12233)

**Plate 17**

2  Trail of sarsen stones brought to their present position by solifluction, Fyfield, near Marlborough. (A.11411)

patches of Clay-with-flints resting on its surface presumably have a long and complex history (Figure 34).

Around the margins of this peneplane there is a marked change of slope below which there is some evidence of planation at heights varying from 198 m above OD in the north and west to 168 m in the south and east. It has long been considered possible that the change in slope marks the landward extent of a late Pliocene–early Pleistocene sea, a view summarised by Wooldridge and Linton (1955). Such a transgression has been established in south-east England by the presence of Pliocene marine deposits at about 180 m above OD at Lenham, Kent, and by early Pleistocene marine deposits at a similar height at Netley Heath, Surrey.

It has been claimed (Everard, 1954) that successive stages in the retreat of the sea are marked in the Hampshire Basin by marine gravels, which can be distinguished from river gravels since their outcrops maintain constant heights above Ordnance Datum (the 'horizontal segments' of some authors). Everard believed that there were level 'terraces' of this sort at five different heights down to 112.5 m above OD, this lowest level approximating to the 'Sicilian terrace' of Bury (1933). By at least the culmination of the earliest major European glaciation, sea level may have fallen below its present position, and so exposed the entire region to subaerial denudation, though Wooldridge and Linton (1955) took a different view. No lower retreat benches of this early age have, however, been identified with certainty. If any do exist their identification is hindered by the repeated readvances of the sea that are likely to have covered the lower ground during interglacial times. The dating of the high beaches is uncertain but, at a height close below that of the 112.5 m 'beach', a fissure-deposit in the chalk at Dewlish, near Dorchester, has yielded bones and teeth of the southern elephant *Archidiskodon meridionalis*, which is characteristic of the Cromerian interglacial, and so appears to set a minimum age for the 112.5 m sea level. This is the oldest Pleistocene deposit in the region to which an age can be assigned on palaeontological grounds. The Plateau Gravels have, as yet, been insufficiently analysed to determine throughout the region which of them grade to these high sea levels.

At lower levels several interglacial marine beaches, each with lower gravels grading to them, are likely to be present. Many terraces have indeed been identified and some of these are linked to horizontal segments. Thus the 'Ambersham terraces' and the 'Sleight terrace' in the Bournemouth area slope from about 70 m at Wimborne to 57m at Sway, and seem to be related to a degraded bluff, which can be traced at a nearly constant level from Stokeford Heath, between Wool and Wareham, across Dorset and Hampshire into the so-called '100 ft raised beach'—the Slindon Raised Beach—in Sussex. The gravel at the foot of this bluff is taken to be marine by Everard; it tops the cliffs at Barton-on-Sea, near Christchurch, and is nearly everywhere overlain by brickearth. The gravel shows periglacial involutions at many localities and contains a fossil ice-wedge at Barton (Lewin, 1966). These frost structures, together with the overlying brickearth and, in places, a drape of Coombe Rock, point to a climatic deterioration after the 'terrace' formed. The Goodwood Raised Beach, at a slightly lower elevation, is similarly buried beneath Coombe Rock. These beaches have commonly been assumed to have formed during the Hoxnian interglacial, but this correlation is far from established.

**Table 13** Stratigraphy and suggested chronology of the Pleistocene in the Hampshire Basin

| Stage | Climate | Events | Years before present |
|---|---|---|---|
| Flandrian | Temperate | Sea level rise of about 45 m disrupting Solent river system and breaching ridge between Isles of Wight and Purbeck. Poole Harbour and Christchurch Bay formed. Low-lying valley mouths drowned | 10 000 |
| Devensian (Weichsel) | Periglacial | Head and Coombe Rock formed. Fisherton brickearth. Sea level at times 18 m below OD. Submerged river terraces | about 70 000 |
| Ipswichian (Eemian) | Temperate | ?Raised beaches at Portland Bill and Selsey; ?freshwater and estuarine deposit at Stone. Sea level about 7.5 m above OD. Solent river system disrupted | ?100 000 |
| Wolstonian | Glacial | Discontinuous local permafrost. Contortions and ice-wedges formed in Hoxnian gravels. Erratics from Devon and Cornwall in Portland raised beach | |
| Hoxnian | Temperate | ?Slindon raised beach and 44 m marine bluff formed. Solent river system disrupted | |
| Anglian | Glacial | 'Solent River' formed as an integrated system. Inland terraces at 70 m and 56.5 m formed. Erratics from Brittany (etc.) dumped by glacier around Selsey | |
| Cromerian | Temperate | **Dewlish fissure deposit with 'southern elephant'** | ?350 000 |
| Beestonian | Cold | Five marine stillstands between about 183 m and 112.5 m ?Striated sarsens and flints at 90 m around Wimborne. ?Some high-level plateau gravels and Clay-with-flints formed | |
| Earlier Pleistocene and Pliocene | | Supposed marine transgression. Planation at ± 183 m | |

Topographic bluffs also bound the lower beaches, most prominent of which are the '50 ft' beach (Portland Raised Beach) and the '15 ft' beach (Selsey Raised Beach) identified around Portsmouth. The former is of interest in containing erratics derived from Cornwall and Devon, possibly transported there by sea-ice, while frost-heave structures in the Purbeck limestones underlying the beach were attributed by Pugh and Shearman (1967) to discontinuous permafrost. The conventional view (West, 1972) is that the Selsey Raised Beach is of Ipswichian age, and it has also been claimed that the Portland Raised Beach is of the same general date. The evidence for this correlation is, however, far from convincing.

Separating the gravels that may have a marine origin from those that are river terraces is not easy, particularly for the older deposits which have been affected by several phases of permafrost and are masked in places by solifluction deposits. The lithologies are not distinctive. Apart from one instance of beach-battered shingle at a high level in the Isle of Wight, all the gravels of the terraces, whether considered marine or fluviatile, consist of angular and sub-angular flints. This, however, is no proof that they are not of marine origin, for the same is true of the gravels now forming in the beaches of Southampton Water and the Solent, which are derived from the surrounding low cliffs and accumulate in water so sheltered that little or no rounding takes place.

While the detailed chronology of the river terrace deposits is thus uncertain, their distribution points strongly to the presence of a 'Solent River' (Figure 35), which initially integrated the consequent streams flowing from north and south at a time when the Isle of Wight was still joined to the mainland. This river was presumably periodically disrupted by the many fluctuations in sea level which also ultimately caused the separation of the Isle of Wight from the mainland after the chalk ridge between Studland, in the Isle of Purbeck, and the Needles was breached. During cold periods, when ice-caps built up in more northern areas, the rivers attempted to grade to the resultant low sea levels; many of their alluvial flats are consequently underlain by buried channels, many of which may have a complex history. The only deposits known in them, however, are those of the Flandrian, and have been built up as the rivers have aggraded to the present sea level.

Away from the streams the most characteristic landform is the chalk downland of Wessex. The typical rounded flowing form of this countryside has led to the suggestion (Te Punga, 1957) that the whole landscape has been moulded under the influence of permafrost and other periglacial processes. Williams (1968) estimated that up to 10.5 m of strata might have been removed by periglacial action on chalk slopes as gentle as $1°$ to $3°$, while solifluction sheets could have moved wholesale down the hillsides. Others (e.g. Small, 1961), however, claim that normal stream erosion and spring sapping can of themselves account for the mass movement that has undoubtedly taken place and can have produced the characteristic landscape without the necessity for periglacial activity. However this may be, the presence of several generations of Head shows that solifluction has been active up to at least the late-Devensian glaciation, some 18 000 years ago; a flood-brickearth at Fisherton, Salisbury, has yielded a mammal fauna of this period. Whatever the respective dominance of these different processes, they have bequeathed to us one of the most beautiful regions of Britain.

Pleistocene and Recent 131

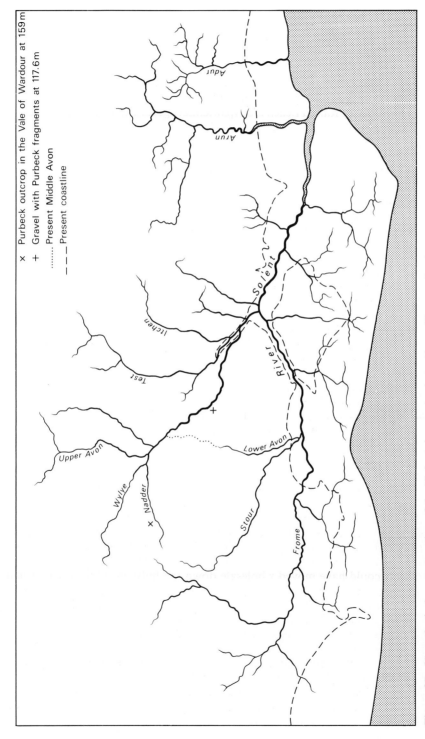

**Figure 35** Hypothetical reconstruction of the 'Solent river system' at a time of low sea level (After Reid, 1902)

# 8 Selected References

The following references, selected from about 1000 that have been examined, are chosen either because they are classic works summarising earlier research, or because they are the latest works seen in particular fields of research. They all give further references which will be helpful to any reader who wishes to pursue a particular topic.

ALLEN, P. 1972. Wealden detrital tourmaline: implications for northwestern Europe. *J. Geol. Soc. London*, Vol. 128, Pt. 3, pp. 273 – 284.
ANDERSON, F.W. 1967. Ostracods from the Weald Clay of England. *Bull. Geol. Surv. G.B.*, No. 27, pp. 237 – 269.
ANDREWS, W.R. and JUKES-BROWNE, A.J. 1894. The Purbeck Beds of the Vale of Wardour. *Q. J. Geol. Soc. London*, Vol. 50, Pt. 1, pp. 44 – 71.
ARKELL, W.J. 1933. *The Jurassic System in Great Britain.* (Oxford: Clarendon Press.)
—1936 Analysis of the Mesozoic and Cainozoic folding in England. *Rep. Int. Geol. Congr. XVI, USA*, No. 2, p. 937.
—1947. The geology of the country around Weymouth, Swanage, Corfe and Lulworth. *Mem. Geol. Surv. G.B.* 386 pp.
BADEN-POWELL, D.F.W. 1930. On the geological evolution of the Chesil Bank. *Geol. Mag.*, Vol. 67, No. 797, pp. 499 – 513.
BLONDEAU, A. and POMEROL, CH. 1969. A contribution to the sedimentological study of the Palaeogene of England. *Proc. Geol. Assoc.*, Vol. 79 (for 1968), Pt. 4, pp. 441 – 455.
BRISTOW, H.W. 1857. Comparative sections of the Purbeck strata of Dorset. Vertical Sections, Sheet 22. *Geol. Surv. England and Wales.*
BROOKFIELD, M.E. 1973. The palaeoenvironment of the Abbotsbury Ironstone. *Palaeontology*, Vol. 16, Pt. 2, pp. 261 – 274.
—1978. The lithostratigraphy of the Upper Oxfordian and lower Kimmeridgian beds of South Dorset, England. *Proc. Geol. Assoc.*, Vol. 89, Pt. 1, pp. 1 – 32.
BUCKMAN, S.S. 1922. Jurassic chronology: II - Preliminary studies. Certain Jurassic strata near Eypesmouth (Dorset): the Junction-Bed of Watton Cliff and associated rocks. *Q. J. Geol. Soc. London*, Vol. 78, Pt. 4, pp. 378 – 457.
BUJAK, J.P., DOWNIE, C., EATON, G.L. and WILLIAMS, G.L. 1980 Dinoflagellate cysts and Acritarchs from the Eocene of Southern England. *Spec. Pap. Palaeontol.*, No. 24, 100 pp.
BURTON, E. ST J. 1933. Faunal horizons of the Barton Beds in Hampshire. *Proc. Geol. Assoc.*, Vol. 44, Pt. 2, pp. 131 – 167.
BURY, H. 1933. The Plateau Gravels of the Bournemouth area. *Proc. Geol. Assoc.*, Vol. 44, pp. 314 – 335.
CARR, A.P. and BLACKLEY, M.W.L. 1973. Investigations bearing on the age and development of Chesil Beach, Dorset, and the associated area. *Trans. Inst. Br. Geogr.*, No. 58, pp. 99 – 111.
CARRECK, J.N. 1955. The Quaternary vertebrates of Dorset, fossil and sub-fossil. *Proc. Dorset Nat. Hist. and Archaeol. Soc.*, Vol. 75, pp. 164 – 188.
CARTER, D.J. and HART, M.B. 1977. Aspects of mid-Cretaceous stratigraphical micropalaeontology. *Bull. Br. Mus. (Nat. Hist.), Geol.*, Vol. 29, No. 1, pp. 1 – 135.

CASEY, R. 1960—. A monograph of the Ammonoidea of the Lower Greensand. Parts I–VII, pp. 1–582, 1–97 (incomplete). *Palaeontogr. Soc.*
— 1961. The stratigraphical palaeontology of the Lower Greensand. *Palaeontology*, Vol. 3, Pt. 4, pp. 487–621.
— 1963. The dawn of the Cretaceous period in Britain. *Bull. South-Eastern Union Sci. Socs.*, No. 117, 15 pp. (Preprint).
— 1967. The position of the Middle Volgian in the British Jurassic. *Proc. Geol. Soc. London*, No. 1640, pp. 128–133.
— 1971. Facies, faunas and tectonics in late Jurassic – early Cretaceous Britain. Pp. 153–168 in *Faunal provinces in space and time*. MIDDLEMISS, F.A., RAWSON, P.F. and NEWALL, G. (Editors). *Spec. Issue Geol. J.*, No. 4. (Liverpool: Seel House Press.)
— 1973. The ammonite succession at the Jurassic – Cretaceous boundary in eastern England. Pp. 193–266 in *The boreal Lower Cretaceous*. CASEY, R. and RAWSON, P.F. (Editors). *Spec. Issue Geol. J.*, No. 5, (Liverpool: Seel House Press.)
— and BRISTOW, C.R. 1964. Notes on some ferruginous strata in Buckinghamshire and Wiltshire. *Geol. Mag.*, Vol. 101, No. 2, pp. 116–128.
CAVELIER, C. 1976. La limite Eocène-Oligocène en Europe occidentale. *Bull. d'Inf. Géol. Bassin de Paris*, Vol. 13, No. 1, pp. 65–84.
CHANDLER, M.E.J. 1964. *The lower Tertiary floras of southern England.* IV. A summary and survey of findings in the light of recent botanical observations. 151 pp. (London: Trustees of the British Museum (Natural History).)
CHATWIN, C.P. 1927. Appendix IX. Fossils from the ironsands on Netley Heath (Surrey). *Summ. Prog. Geol. Surv. G.B. for 1926*, pp. 154–157.
— 1960. *British Regional Geology. The Hampshire Basin and adjoining areas.* Third Edition, 100 pp. (London: HMSO.)
— 1963. Palaeontology of the Lenham Beds. Pp. 91–100 in WORSSAM, B.C. Geology of the country around Maidstone. *Mem. Geol. Surv. G.B.*
COOPER, J. 1976. British Tertiary stratigraphical and rock terms, formal and informal, additional to Curry, 1958, *Lexique stratigraphique international*. *Spec. Pap. Tertiary Research*, No. 1, 37 pp.
COPE, J.C.W. 1967. The palaeontology and stratigraphy of the lower part of the upper Kimmeridge Clay of Dorset. *Bull. Br. Mus. (Nat. Hist.), Geol.*, Vol. 15, No. 1, pp. 1–79.
— 1978. The ammonite faunas and stratigraphy of the upper part of the Upper Kimmeridge Clay of Dorset. *Palaeontology*, Vol. 21, Pt. 3, pp. 469–533.
— and WIMBLEDON, W.A. 1973. Ammonite faunas of the uppermost Kimmerridge Clay, the Portland Sand and the Portland Stone of Dorset. *Proc. Ussher Soc.*, Vol. 2, Pt. 6, pp. 593–598.
COSTA, C.I. and DOWNIE, C. 1976. The distribution of the dinoflagellate *Wetzeliella* in the Palaeogene of north-western Europe. *Palaeontology*, Vol. 19, Pt. 4, pp. 591–614.
— and EATON, G.L. 1976. Palynostratigraphy of some Middle Eocene sections from the Hampshire Basin (England). *Proc. Geol. Assoc.*, Vol. 87, Pt. 3, pp. 273–284.
COX, B.M. and GALLOIS, R.W. 1981. The stratigraphy of the Kimmeridge Clay of the Dorset type area and its correlation with some other Kimmeridgian sequences. *Rep. Inst. Geol. Sci.*, No. 80/4, 44 pp.
COX, L.R. 1929. A synopsis of the Lamellibranchia and Gastropoda of the Portland Beds of England. *Proc. Dorset Nat. Hist. and Archaeol. Soc.*, Vol. 50, pp. 130–202, pls., refs.

CURRY, D. 1958. *Lexique stratigraphique international.* Europe, Fascicule 3a, England, Wales and Scotland, Part 3a XII, Palaeogene. 82 pp. (Paris: Centre National de la Recherche Scientifique.)
— 1965. The Palaeogene beds of south-east England. *Proc. Geol. Assoc.,* Vol. 76, Pt. 2, pp. 151–173.
— 1976. The age of the Hengistbury Beds (Eocene) and its significance for the structure of the area around Christchurch, Dorset. *Proc. Geol. Assoc.,* Vol. 87, pp. 401–407.
— KING, A.D., KING, C. and STINTON, P.C. 1977. The Bracklesham Beds (Eocene) of Bracklesham Bay and Selsey, Sussex. *Proc. Geol. Assoc.,* Vol. 88, Pt. 4, pp. 243–254.
— and SMITH, A.J. 1975. The structure and geological evolution of the English Channel. *Philos. Trans. R. Soc. London,* Ser.A., Vol. 279, pp. 3–20.
— and WISDEN, D.E. 1958. Geology of some British coastal areas: The Southampton district, including Barton (Hampshire) and Bracklesham (Sussex) coastal sections. *Geol. Assoc. Guides* No. 14.
DALEY, B. 1971. Diapiric and other deformational structures in an Oligocene argillaceous limestone. *Sediment. Geol.,* Vol. 6, No. 1, pp. 29–51.
— 1972. Some problems concerning the early Tertiary climate of southern Britain. *Palaeogeogr., Palaeoclimatol., Palaeoecol.,* Vol. 11, No. 3, pp. 177–190.
— 1973. The palaeoenvironment of the Bembridge Marls (Oligocene) of the Isle of Wight, Hampshire. *Proc. Geol. Assoc.,* Vol. 84, Pt. 1, pp. 83–93.
— and EDWARDS, N. 1971. Palaeogene warping in the Isle of Wight. *Geol. Mag.,* Vol. 108, pp. 399–405.
DAVIES, D.K. 1967. Origin of friable sandstone–calcareous sandstone rhythms in the Upper Lias of England. *J. Sediment. Petrol.,* Vol. 37, No. 4, pp. 1179–1188.
— 1969. Shelf sedimentation: an example from the Jurassic of Britain. *J. Sediment. Petrol.,* Vol. 39, No. 4, pp. 1344–1370.
DELAIR, J.B. 1958–1960. The Mesozoic reptiles of Dorset. *Proc. Dorset Nat. Hist. and Archaeol. Soc.,* Pt. 1, Vol. 79, pp. 47–72, 1958; Pt. 2, Vol. 80, pp. 52–90, 1959; Pt. 3, Vol. 81, pp. 59–85, 1960.
DIKE, E.F. 1972. *Ophiomorpha nodosa* Lundgren: environmental implications in the Lower Greensand of the Isle of Wight. *Proc. Geol. Assoc.,* Vol. 83, pp. 165–177.
DINES, H.G. and CHATWIN, C.P. 1930. Pliocene sands from Rothamsted (Hertfordshire). *Summ. Prog. Geol. Surv. G.B. for 1929,* Pt.III, pp. 1–7.
DOUGLAS, J.A. and ARKELL, W.J. 1928. The stratigraphical distribution of the Cornbrash: I. The south-western area. *Q.J. Geol. Soc. London,* Vol. 84, Pt. 1, pp. 117–178.
DRUMMOND, P.V.O. 1970. The mid-Dorset swell. Evidence of Albian–Cenomanian movements in Wessex. *Proc. Geol. Assoc.,* Vol. 81, pp. 679–714.
EATON, G.L. 1976. Dinoflagellate cysts from the Bracklesham Beds (Eocene) of the Isle of Wight, southern England. *Bull. Br. Mus. (Nat. Hist.), Geol.,* Vol. 26, No. 6, pp. 227–332.
EVERARD, C.E. 1954. The Solent river: a geomorphological study. *Inst. Br. Geogr.,* Publ. No. 20, *Trans. and Papers* 1954, pp. 41–58.
FISHER, O. 1862. On the Bracklesham Beds of the Isle of Wight Basin. *Q. J. Geol. Soc. London,* Vol. 18, *Proc.,* pp. 65–94.
FUERSICH, F.T. 1976. Fauna-substrate relationships in the Corallian of England and Normandy. *Lethaia,* Vol. 9, pp. 343–356.
— 1977. Corallian (Upper Jurassic) marine benthic associations from England and Normandy. *Palaeontology,* Vol. 20, Pt. 2, pp. 337–385. figs., refs.

GALTON, P.M. 1974. The Ornithischian Dinosaur *Hypsilophodon* from the Wealden of the Isle of Wight. *Bull. Br. Mus. (Nat. Hist.), Geol.*, Vol. 25, No. 1, pp. 1 – 152c.
GARDNER, J.S. 1882. Description and correlation of the Bournemouth Beds. Part II. Lower or Freshwater Series. *Q.J. Geol. Soc. London*, Vol. 38, pp. 1 – 15.
GILKES, R.J. 1978. On the clay mineralogy of upper Eocene and Oligocene sediments in the Hampshire Basin. *Proc. Geol. Assoc.*, Vol. 89, Pt. 1, pp. 43 – 56.
GREEN, J.F.N. 1946. The terraces of Bournemouth, Hants. *Proc. Geol. Assoc.*, Vol. 57, Pt. 2, pp. 82 – 101.
— 1947. Some gravels and gravel-pits in Hampshire and Dorset. *Proc. Geol. Assoc.*, Vol. 58, Pt. 2, pp. 128 – 143.
GRUAS-CAVAGNETTO, C. 1976. Etude palynologique du Palaéogène du sud de l'Angleterre. *Cahiers de Micropaléont.*, No. 1, 50 pp., refs., pls.
HALLAM, A. 1961. Cyclothems, transgressions and faunal change in the Lias of north-west Europe. *Trans. Edinburgh Geol. Soc.*, Vol. 18, Pt. 2, pp. 124 – 173.
— 1964. Origin of the limestone-shale rhythm in the Blue Lias of England: a composite theory. *J. Geol.*, Vol. 72, No. 2, pp. 157 – 169.
— 1969. A pyritised limestone hardground in the Lower Jurassic of Dorset. *Sedimentology*, Vol. 12, Nos. 3 – 4, pp. 231 – 240.
— 1970. *Gyrochorte* and other trace fossils in the Forest Marble (Bathonian) of Dorset, England. Pp. 189 – 200 in *Trace Fossils*. CRIMES, T.P. and HARPER, J.C. (Editors). *Spec. Issue J. Geol.*, No. 3. (Liverpool: Seel House Press.)
— and SELLWOOD, B.W. 1976. Middle Mesozoic sedimentation in relation to tectonics in the British area. *J. Geol.*, Vol. 84, pp. 301 – 321.
HANCOCK, J.M. 1969. Transgression of the Cretaceous sea in south-west England. *Proc. Ussher Soc.*, Vol. 2, Pt. 2, pp. 61 – 83.
— and CURRY, D. 1977. Excursion à l'Ile de Wight. *Bull. Inform. Géol. Bassin de Paris*, Vol. 14, No. 3, pp. 5 – 29.
HAWKINS, A.B. and KELLAWAY, G.A. 1971. Field meeting at Bristol and Bath with special reference to new evidence of glaciation. *Proc. Geol. Assoc.*, Vol. 82, Pt. 2, pp. 267 – 292.
HOOKER, J.J. 1975. Report of field meeting to Barton, Hampshire, October 12th, 1974. *Tertiary Times*, Vol. 2, pp. 163 – 167.
— and INSOLE, A.N. 1980. The distribution of mammals in the English Palaeogene. *Tertiary Research.*, Vol. 3, Pt. 1, pp. 31 – 45.
HOUNSELL, S.S.B. 1952. Portland and its stone. *Mine Quarry Eng.*, Vol.18, No. 4, pp. 107 – 114.
HOUSE, M. R. 1961. Structure of the Weymouth Anticline. *Proc. Geol. Assoc.*, Vol.72, pp.221 – 238.
JACKSON, J.F. 1926. The Junction-Bed of the Middle and Upper Lias on the Dorset coast. *Q. J. Geol. Soc. London*, Vol. 82, Pt. 4, pp. 490 – 525.
JACQUÉ, M. and THOUVENIN, J. 1975. Lower Tertiary tuffs and volcanic activity in the North Sea. Pp. 455 – 465 in *Petroleum and the continental shelf of north-west Europe*. WOODLAND, A.W. (Editor). (Barking: Applied Science Publishers.)
JAMES, J.P. 1974. Report of field meeting to Lower Swanwick, Hampshire. *Tertiary Times*, Vol. 2, Pt. 1, pp. 23 – 28; Pt. 2, p. 72.
JEANS, C.V., MERRIMAN, R.J. and MITCHELL, J.G. 1977. Origin of the Middle Jurassic and Lower Cretaceous fuller's earths in England. *Clay Miner.*, Vol. 12, pp. 11 – 44.
JENKYNS, H.C. and SENIOR, J. 1977. A Liassic palaeofault from Dorset. *Geol. Mag.*, Vol. 114, No. 1, pp. 47 – 52.

JUKES-BROWNE, A.J. and SCANES, J. 1901. On the Upper Greensand and Chloritic Marl of Mere and Maiden Bradley in Wiltshire. *Q.J. Geol. Soc. London*, Vol. 57, pp. 96–125.

KEEN, M.C. 1977. Ostracod assemblages and the depositional environments of the Headon, Osborne and Bembridge Beds (Upper Eocene) of the Hampshire Basin. *Palaeontology*, Vol. 20, Pt. 2, pp. 405–445.

KELLAWAY, G.A., REDDING, J.H., SHEPHARD-THORN, E.R. and DESTOMBES, J.-P. 1975. The Quaternary history of the English Channel. *Philos. Trans. R. Soc. London*, Ser. A, Vol. 279, pp. 189–218.

KENT, P.E. 1949. A structure contour map of the surface of the buried pre-Permian rocks of England and Wales. *Proc. Geol. Assoc.*, Vol. 60, Pt. 2, pp. 87–104.

KENNEDY, W.J. 1970. A correlation of the uppermost Albian and the Cenomanian of south-west England. *Proc. Geol. Assoc.*, Vol. 81, Pt. 4, pp. 613–677.

— and GARRISON, R.E. 1975. Morphology and genesis of nodular chalks and hardgrounds in the Upper Cretaceous of southern Europe. *Sedimentology*, Vol. 22, No. 3, pp. 311–386.

KING, C. 1981. The stratigraphy of the London Clay and associated deposits. *Spec. Pap. Tertiary Research*, No. 6. Tertiary Research Group. 158 pp.

LAING, J.F. 1975. Mid-Cretaceous angiosperm pollen from southern England and northern France. *Palaeontology*, Vol. 18, Pt. 4, pp. 775–808.

LEWIN, J. 1966. Fossil ice-wedges in Hampshire. *Nature*, Vol. 211, No. 5050, p. 728, refs.

MACFADYEN, W.A. 1970. *Geological highlights of the west country.* 296 pp. (London: Butterworth.)

MURRAY, J.W. and WRIGHT, C.A. 1974. Palaeogene Foraminiferida and palaeoecology, Hampshire and Paris basins and the English Channel. *Spec. Pap. Palaeontol.*, No. 14, 177 pp.

OLDHAM, T.C.B. 1976. Flora of the Wealden plant debris beds of England. *Palaeontology*, Vol. 19, Pt. 3, pp. 437–502.

OWEN, H.G. 1971. Middle Albian stratigraphy in the Anglo-Paris Basin. *Bull. Br. Mus. (Nat. Hist.), Geol.*, Suppl. 8, 164 pp.

PALMER, C.P. 1966a. Note on the fauna of the Margaritatus Clay (Blue Band) in the Domerian of the Dorset coast. *Proc. Dorset Nat. Hist. and Archaeol. Soc.*, Vol. 87, pp. 67–68.

— 1966b. The fauna of Day's Shell Bed in the Middle Lias of the Dorset coast. *Proc. Dorset Nat. Hist. and Archaeol. Soc.*, Vol. 87, pp. 69–80.

PALMER, J.T. and FUERSICH, F.T. 1974. The ecology of a Middle Jurassic hardground and crevice fauna. *Palaeontology*, Vol. 17, Pt. 3, pp. 507–524.

PARSONS, C.F. 1976. A stratigraphic revision of the *humphriesianum/subfurcatum* Zone rocks (Bajocian Stage, Middle Jurassic) of southern England. *Newsl. Stratigr.*, Vol. 5, Nos. 2/3, pp. 114-142.

PENN, I.E., MERRIMAN, R.J. and WYATT, R.J. 1979. A proposed type section for the Fuller's Earth (Bathonian), based on the Horsecombe Vale No. 15 Borehole, near Bath, and including details of contiguous strata. Part I in The Bathonian strata of the Bath–Frome area. *Rep. Inst. Geol. Sci.*, No. 78/22.

— and WYATT, R.J. 1979. The stratigraphy and correlation of the Bathonian strata in the Bath–Frome area. Part 2 in The Bathonian strata of the Bath–Frome area. *Rep. Inst. Geol. Sci.*, No. 78/22.

PRESTWICH, J. 1846. On the Tertiary or Supracretaceous Formations of the Isle of Wight, etc. *Q. J. Geol. Soc. London*, Vol. 2, pp. 255–259.

PUGH, M.E. and SHEARMAN, D.J. 1967. Cryoturbation structures at the south end of the Isle of Portland. *Proc. Geol. Assoc.*, Vol. 78, Pt. 3, pp. 403–471.

RAWSON, P.F. and others. 1978. A correlation of Cretaceous rocks in the British Isles. *Spec. Rep. Geol. Soc. London.* No. 9, 70 pp.
REID, C. 1902. The geology of the country around Ringwood. *Mem. Geol. Surv. G.B.* 62 pp.
RICHARDSON, L. 1928. The Inferior (Oolite) and contiguous deposits of the Burton Bradstock–Broadwindsor district, Dorset. *Proc. Cotteswold Nat. Field Club,* Vol. 23, pp. 35–68.
RIDD, M.F. 1973. The Sutton Poyntz, Poxwell and Chaldon anticlines, southern England: a reinterpretation. *Proc. Geol. Assoc.,* Vol. 84, pp. 1–8.
SELLWOOD, B.W. 1970. The relation of trace fossils to small-scale sedimentary cycles in the British Isles. Pp. 489–504 in *Trace Fossils.* CRIMES, T.P. and HARPER, J.C. (Editors). *Spec. Issue Geol. J.,* No. 3. (Liverpool: Seel House Press.)
— DURKIN, M.A. and KENNEDY, W.J. 1970. Field meeting on the Jurassic and Cretaceous rocks of Wessex. *Proc. Geol. Assoc.,* Vol. 81, Pt. 4, pp. 715–732.
— and JENKYNS, H.C. 1975. Basins and swells and the evolution of an epeiric sea (Pliensbachian–Bajocian of Great Britain). *J. Geol. Soc. London,* Vol. 131, Pt. 4, pp. 373–388.
SMALL, R.J. 1961. The morphology of Chalk escarpments: a critical review. *Inst. Br. Geogr.,* Publ. No. 29, *Trans. and Papers* 1961, pp. 71–90.
SPARKS, B.W. and WEST, R.G. 1972. *The Ice Age in Britain.* 302 pp. (London: Methuen.)
STINTON, F.W. 1970. Field meeting in the New Forest, Hants. *Proc. Geol. Assoc.,* Vol. 81, Pt. 2, pp. 269–274.
— 1975. Fish otoliths from the English Eocene, Pt. 1. *Palaeontogr. Soc.,* pp. 1–56.
STRAHAN, A. 1898. The geology of the Isle of Purbeck and Weymouth. *Mem. Geol. Surv. England and Wales,* 278 pp.
SYKES, R.M. and SURLYK, F. 1976. A revised ammonite zonation of the Boreal Oxfordian and its application in northeast Greenland. *Lethaia,* Vol. 9, pp. 421–436.
TAITT, A.H. and KENT, P.E. 1939. Notes on an examination of the Poxwell anticline. *Geol. Mag.,* Vol. 76, pp. 173–181.
TALBOT, M.R. 1973. Major sedimentary cycles in the Corallian Beds (Oxfordian) of southern England. *Palaeogeogr., Palaeoclimatol., Palaeoecol.,* Vol. 14, No. 4, pp. 293–313.
TE PUNGA, M.T. 1957. Periglaciation in southern England. *Tijdschr. K. Ned. Aardrijkskd. Genoot.,* Vol. 74, pp. 400–412.
TORRENS, H. (Editor) 1969. *International field symposium on the British Jurassic.* 71 pp. (Keele: Keele University.)
TOWNSON, W.G. 1975. Lithostratigraphy and deposition of the type Portlandian. *J. Geol. Soc. London,* Vol. 131, Pt. 6, pp. 619–638.
TURNER, C. 1975. P. 256 in Discussion of DESTOMBES, J.-P. SHEPHARD-THORN, E.R. and REDDING, J.H., A buried valley system in the Strait of Dover. *Philos. Trans. R. Soc. London,* Ser. A, Vol. 279, pp 243–256.
VENABLES, E.M. 1963. The London Clay of Bognor Regis. *Proc. Geol. Assoc.,* Vol. 73 (for 1962), Pt. 3, pp. 245–272.
WEST, I.M. 1960. On the occurrence of celestine in the Caps and Broken Beds at Durlston Head, Dorset. *Proc. Geol. Assoc.,* Vol. 71, Pt. 4, pp. 391–401.
— 1964. Deformation of the incompetent folds in the Purbeck Anticline. *Geol. Mag.,* Vol. 101, p.373.
— 1975. Evaporites and associated sediments of the basal Purbeck Formation (Upper Jurassic) of Dorset. *Proc. Geol. Assoc.,* Vol. 66, Pt. 2, pp. 205–225.

WEST, R.G. 1972. Relative land-sea-level changes in southeastern England during the Pleistocene. *Philos. Trans. R. Soc. London*, Ser. A., Vol. 272, pp. 87–98.

WHITAKER, W. and EDWARDS, W.N. 1926. Wells and springs of Dorset. *Mem. Geol. Surv. G.B.* 119 pp.

WHITE, H.J.O. 1921. Short account of the geology of the Isle of Wight. *Mem. Geol. Surv. G.B.* 201 pp

WHITE, P.H.N. 1948. Gravity data obtained in Great Britain by the Anglo-American Oil Co. Ltd. *Q. J. Geol. Soc. London*, Vol. 104, pp. 339–364.

WHITTAKER, A. 1972. The Watchet Fault — a post-Liassic transcurrent reverse fault. *Bull. Geol. Surv. G.B.*, No. 41, pp. 75–80.

— 1975. A postulated post-Hercynian rift valley system in southern Britain. *Geol. Mag.*, Vol. 112, pp. 137–149.

WILLIAMS, R.B.G. 1965. Permafrost in England during the last glacial period. *Nature, London*, Vol. 205, No. 4978, pp. 1304–1305.

— 1968. Some estimates of periglacial erosion in southern and eastern England. *Biul. Peryglac.*, No. 17, pp. 311–335.

WILSON, R.C.L. 1968a. Upper Oxfordian palaeogeography of southern England. *Palaeogeogr., Palaeoclimatol., Palaeoecol.*, Vol. 4, No. 1, pp. 5–28.

— 1968b. Carbonate facies variation within the Osmington Oolite Series in southern England. *Palaeogeogr., Palaeoclimatol., Palaeoecol.*, Vol. 4, No. 2, pp. 89–123.

WILSON, V., WELCH, F.B.A., ROBBIE, J.A. and GREEN, G.W. 1958. Geology of the country around Bridport and Yeovil. *Mem. Geol. Surv. G.B.* 239 pp.

WIMBLEDON, W.A. 1976. The Portland Beds (Upper Jurassic) of Wiltshire. *Wiltshire Nat. Hist. Mag.*, Vol. 71, pp. 3–11.

— and COPE, J.C.W. 1978. The ammonite faunas of the English Portland Beds and the zones of the Portlandian Stage. *J. Geol. Soc. London,* Vol. 135, Pt. 2, pp. 183–190.

WOODWARD, H.B. 1893. The Jurassic rocks of Britain, Vol. 3, Lias of England and Wales. *Mem. Geol. Surv. G.B.*

WOOLDRIDGE, S.W. and LINTON, D.L. 1955. *Structure, surface and drainage in south-east England.* 176 pp. (London: Philip.)

WRIGHT, C.A. 1972a. Foraminiferids from the London Clay of Lower Swanwick and their palaeoecological interpretation. *Proc. Geol. Assoc.*, Vol. 83, Pt. 3, pp. 337–347.

— 1972b. The recognition of a planktonic foraminiferid datum in the London Clay of the Hampshire Basin. *Proc. Geol. Assoc.*, Vol. 83, Pt. 4, pp. 413–419.

WRIGHT, C.W. and KENNEDY, W.J. 1981. The Ammonoidea of the Plenus Marls and the Middle Chalk. *Palaeontogr. Soc.*, 148 pp., 32 pls.

WRIGHT, J.K. 1981. The Corallian rocks of north Dorset. *Proc. Geol. Assoc.*, Vol. 92, pp. 17–32.

ZIEGLER, B. 1962. Die Ammoniten-Gattung *Aulacostephanus* im Oberjura (Taxionomie, Stratigraphie, Biologie). *Palaeontographica*, Abt. A, Vol. 119, Lief. 1–4, pp. 1–172.

# Index

Abbotsbury  38, 40, 102, 126
Abbotsbury Fault  114
Abbotsbury Ironstone  38
Abbotsbury Swannery  27
Agglestone Grits  100, 102
Aldwick Beds  95
Algal mats  56
ALLEN, P.  65, 122
Alluvium  124
Alpine Orogeny  112, 114
Alum Bay  102, 105, 108
Ambersham terraces  128
'Ammonitico rosso'  17
Amphipod crustaceans  26
Ancliff Oolite  30
ANDERSON, F.C.  65
ANDREWS, W.R.  64
Anglo-Paris-Belgian Basin  91
Anhydrite  56
Ansty  75
ARKELL, W.J.  27, 32, 43, 48, 53, 60, 100, 102, 114, 121, 126
Arreton  67
Atherfield  67, 72
Atherfield Clay  70, 72
Atherfield coastguard station  70
Auversian Stage  100
Aylesbury  43, 50

BADEN POWELL, D.F.W.  126
Baggridge Hill  28
Bagshot Beds  96
Bajocian  18
Barn Rock Bed  95
Barnes High  65
Barton Clay  105
Barton Formation  105
Barton on Sea  128
Barton Sands  105
'Basebed'  50
Basement Bed  94
Batcombe  79

Bath  23, 27, 30
Bath Oolite  28, 30
Bath Stone  23, 28
Bathonian Stage  25
*baylei* Zone  38
Beaminster  10, 18, 25, 79
Beaminster Fault  114
Becton Member  105
Bed A1  105
Bed A2  105
Bed C  105
Bed E  105
Bed F  105
Bed G  105
Bed H  105
Bed I  105
Bed J  105
Bed K  105
Beer Head  126
Belemnite Marls  14
Belemnite Stone  14
Bembridge  91
Bembridge Beds  90
Bembridge Formation  105, 109
Bembridge Limestone  108
Bembridge Marls  108, 109
Bembridge – St Valery line  121
Bencliff Grit  36, 32
Bere Regis  26
Berkshire Oolites  39
Bicavea Bed  88
*bifrons* Zone  18
Birchi Nodule Bed  13
Black Arnioceras Bed  13
Black Band  110
Black Nore Member  43, 47
Black Ven  10, 13, 74, 75, 79
Black Ven Marls  13
Blackgang  75
Blackgang Chine  71, 72
Blackstone  40
Blandford  67
BLONDEAU, A.  90, 94, 96
Blue Lias  10

## 140  *The Hampshire Basin*

Bognor   95, 96
Bognor Rock   95
Bonchurch   75
Boueti Bed   26, 28
Bouguer Anomaly   112
Bouldnor   109
Bouldnor Cliff   109
Bournemouth   91
Bournemouth Freshwater Beds   100, 102
Box   20
Bracklesham Bay   100, 102
Bracklesham Group   100
Bradford Clay   30
Bradford Coral Bed   30
Bradford on Avon   27
Brading   75
Bramshaw   102
Brickearth   125
Bridport   10, 121
Bridport Fault   114
Bridport Sands   17, 18, 20
BRISTOW, C.R.   62
BRISTOW, H.W.   64
Brittany   122
Brixton Bay   65
Brockenhurst Bed   108
Broken Beds   56, 60
Broken Shell Limestone   62
Brook Chine   65
BROOKFIELD, M.E.   38
Buckland Newton   79
BUCKMAN, S.S.   18, 26
BUJAK, J.P.   93
Burning Cliff   40
Burr   62
Bursledon   95
Burton Bradstock   20
BURTON, E. ST J.   105
BURY, H.   128

Calcaire de Brie   110
Calne   38, 74
Calne Freestone   38
Caps   56
Carstone   70, 71, 74
CARTER, D.J.   86
CASEY, R.   2, 40, 43, 48, 54, 62, 67, 72
Castle Hill   94
CAVALIER, C.   93
Celestite   47
Cenomanian   75
Central Somerset Graben   114
Cephalopod Bed   18
Cerithium Beds   110

Chalcedony   56, 60
Chaldon Pericline   114
Chalk   82
Chalk Basement Bed   79, 80, 84
Chalk Marl   84, 86
Chalk Rock   88
Chalky Series   53
Chama Bed   105
CHANDLER, M.E.J.   96
Chapmans Pool   40
Charmouth   13, 79
Chelborough Fault Belt   79
Chert Beds   79, 80
Cherty Freshwater Beds   56, 60
Cherty Series   48
Chesil Bank   126
Chesters Hill   27
Chickerell   32
Chicksgrove Limestone Member   50, 53
Chideock   20
Chief Beef Beds   62
Chilmark   53
Chilmark Oolite Member   53
Chondrites   8, 14, 83
Christchurch   91, 95, 100, 102
Church Cliffs   10
Cinder Bed   54, 62, 64
Clarendon   96
Clay-with-flints   124
Cliff End   105
Cliff End, Burton Bradstock   26
Cliff End Sand   71
Coccolithophorids   83
Coinstone   14
Colwell Bay   105, 108
Combe Down Oolite   28
Compton Bay   72, 74
Combe Martin Fault   121
Coombe Rock   125, 128
COOPER, J.   93, 94, 100, 102
COPE, J.C.W.   40, 43, 48
Corallian Beds   32, 38
*coranguinum* Zone   88
Corbula Beds   62, 109
*cordatum* Zone   32
Corfe   72
Cornbrash   25, 26
Cornstones   80
Cornubian Hinterland   90
Corralian Beds   30
Corsham Coral bed   30
Corston Beds   30
*cortestudinarium* Zone   88
Corton Hill   47
Corton Hill Member   47, 53

COSTA, C.I.   100
Cothelstone Fault   121, 123
Coulston   53
Cowstones   75, 79
COX, L.R.   48
Crackers   70, 72
Cranborne   94, 96
Cranborne Borehole   114, 122
Crediton Trough   114
Creechbarrow Beds   102
Cretaceous   62
Crewkerne   20
Crockwood   53
Cromerian Interglacial   128
Crook Hill Brickyard   32
Cuisian Stage   100
CURRY, D.   91, 93, 96, 100, 105, 109, 121
Cyclothem   7
*cymodoce* Zone   38
*Cypridea valdensis* Zone   65
'Cypris' Freestones   56, 60

DALEY, B.   96
Dancing Ledge Member   48
DAVIES, D.K.   18, 109, 122
Days Shell Bed   16
Dean Hill Anticline   121, 122
Decalcified till   124
DELAIR, J.B.   62
Devensian Glaciation   130
Devizes   72
Dew Bed   20
Dewlish   128
Dieppe   91
Digona Bed   26, 28
DIKE, E.F.   71
Dinosaurs   62
Dinton   64, 67
*Diplocraterion*   7, 37
Dirt Bed   56
Dorchester   95
DOUGLAS, J.A.   27
Down Cliff Clay   17
Down Cliff Sands   16
DRUMMOND, P.V.O.   79, 123
Dungy Head   40, 47, 50
Dungy Head Member   47, 48
Durdle Door   50
Durdle Promontory   67
Durlston Bay   56, 61, 62, 64
Durlston Beds   54
Durlston Formation   62, 64
Durlston Head   47, 50, 60

Earnley Formation   100

East Basin   47, 48
East Cliff, Burton Bradstock   22
East Fleet   32
East Fleet church   30
EATON, G.L.   96
EDWARDS, N.   122
Eggardon Grit   79
Eggardon Hill   79
EVERARD, C.E.   128
Exogyra Rock   79
Eype Clay   16
Eype Nodule Bed   16

*falciferum* Zone   17
Fareham   95
Faulkland   28
Fawley   102
Ferrugineous Bands   72
Ferrugineous Sands   70
Fish Bed   10
FISHER, O.   100
Fisherton   130
Flandrian   130
Fleet   20, 30, 126
Foliated Clay and Sand   71
*forbesi* Zone   72
Fordingbridge   26, 67, 95, 96
Fordingbridge 'High'   114, 121, 122, 123
Fordingbridge Borehole   122
Forest Marble   23, 26, 27, 28, 30
Foxmould   79
Frome Clay   23, 25, 26, 28
FUERSICH, F.T.   30, 38
Fullers Earth   23, 26
Fullers Earth Rock   25, 26, 27
Furzedown Clays   32
Furzey Cliff   32

Gad Cliff   40, 43, 47
Gad Cliff Member   47, 50, 53
GALTON, P.M.   65
GARRISON, R.E.   82, 83
Gault   67, 74
GILKES, R.J.   90
Glauconitic Marl   79, 80, 84
Golden Cap   14, 16, 74, 79
Goodwood Raised Beach   128
Gosport   102
Graben   112
GRUAS-CAVAGNETTO, C.   96
Great Dirt Bed   60
Great Oolite   20, 28
Great Oolite Limestone   25, 26
Great Oolite Series   23
Green Ammonite Beds   14
Grey Chalk   84, 86

142  *The Hampshire Basin*

Guildford   122
Gypsum   56, 60

Halfway House   22
HALLAM, A.   8, 10, 14, 26, 43
Ham Cliff   32
Ham Hill   18
Ham Hill Stone   18
Hampshire – Dieppe Basin   91
Hamstead   109
Hamstead Formation   105, 108, 109
HANCOCK, J.M.   79, 93, 109
Hard Cockle Beds   56, 60
Hard Marl   10
Hardgrounds   14, 83
Hardy Monument   102
HART, M.B.   86
Haselbury   20
Haydon   25
Head   125
Headon Beds   90
Headon Hill   105, 108
Headon Hill Limestone   108
Headon Member   105, 108
Hengistbury Beds   105
Herbury   26
Hercynian Front   112, 123
Highcliff Member   105
Hinton Charterhouse   28
Hinton Sands   26
Holworth House   40, 48, 75, 79
Hooke   79
Hooke Fault   121
Hordle Cliff   105
Hordle Member   105
'Horizontal Segments'   128
Horse Ledge   71
Horst   112
Houns Tout Cliff   40, 43
HOUNSELL, S.S.B.   50
HOUSE, M.R.   114
How Ledge Limestone   105
Hoxnian Interglacial   128
Hummocky   14
Huntingbridge Formation   100
Huntingbridge Member   102

Ice Wedge   128
*Iguanodon*   65
Inferior Oolite   18, 122
Intermarine Beds   62
Ipswichian   130
Isle of Portland   43, 47, 48, 50
Isle of Purbeck   48, 62, 114
Isle of Purbeck Monocline   121

Isle of Wight   74, 86, 88
Isle of Wight Monocline   114

JACKSON, J.F.   17
JACQUE, M.   96
JAMES, J.P.   95
JENKYNS, H.C.   7, 8, 17
Jordan cliff   32
Jordan cliff clays   32
JUKES-BROWN, A.J.   64, 80
Junction Bed   10, 16, 17, 22
Jurassic   7

KEEN, M.C.   90
KELLAWAY, G.A.   124
Kellaways   30
Kellaways Beds   30
Kellaways Clay   30
Kellaways Rock   30
KENNEDY, W.J.   79, 82, 83
Kennet Valley   75
KENT, P.E.   3, 114
Kimmeridge   39, 26
Kimmeridge Clay   30, 38, 39, 40, 43, 114
Kimmeridge Coal   40
Kingsclere   20, 61, 64, 67, 75
Kingsclere Borehole   122
Knorri Clays   25

Ladder Chine   71
Ladder Sands   71
LAING, J.F.   80
*lamberti* Zone   32
Laminated Beds   16
Langton Herring   25, 26
*lata* Zone   86
Leaf Bed   105
Lenham   128
*levesquei* Zone   17, 18, 20
LEWIN, J.   128
Lias   8
Lingula Shales   43
LINTON, D.L.   128
Litton Cheney   79
London Clay Basement Bed   94
Lopen – Chiselborough – Coker Fault System   114
Lower Building Stones   60, 62
Lower Cenomanian   79, 80
Lower Chalk   84
Lower Clay   95
Lower Cornbrash   25, 26, 27
Lower Crioceras Bed   70
Lower Fullers Earth   27
Lower Greensand   67, 72

Lower Gryphaea Bed 70
Lower Hamstead Beds 110
Lower Headon Beds 105
Lower Lias 10
Lower Lobster Bed 70, 72
Lower Swanwick 95
Luccomb Chine 72
Lulworth 54
Lulworth Banks Borehole 122
Lulworth Beds 54
Lulworth Cove 43, 72
Lulworth Crumple 121
Lulworth Formation 7, 30, 56, 60, 61, 64
Lutetian Stage 100
Lyme Regis 7, 74
Lymington 108
Lyndhurst 108

MACFADYEN, W.A. 126
Maiden Bradley 80
Maiden Newton 79
Main Building Stones 53
Mammal Bed 56, 60
Mappowder 38
Marble Beds 62
Margaritatus Clay 16
Margaritatus Stone 16
*mariae* Zone 32
Marlstone Rock Bed 16, 17
Marly Freshwater Beds 56, 60
Marnhull Freestone 38
Marshwood 10
Marshwood Dome 114
Massive Bed 43
Melbourn Rock 86
Mendip Axis 27
Mere 80
Mere Fault 114
Micaceous Beds 16
Mid-Dorset Swell 79, 123
Middle Albian 75
Middle Cenomanian 79
Middle Chalk 86
Middle Clay 95
Middle Inferior Oolite 20
Middle Jurassic 18
Middle Lias 16
Middle Lias Marls 16
Midford Sands 18
Milborne Port 27
Milford 105
Miocene 126
*mucronata* Zone 88, 90
Mudeford 105

Mupe Rocks 50
Mupe Bay 61, 64, 67
MURRAY, J.W. 95, 102

Naish Member 105
Netley Heath 128
Nettlestone Grits 108
New Forest 100, 102
New Red Sandstone 114, 122
North Coker 20
North Sea Basin 122
Nothe Clay 32, 36, 39
Nothe Grit 32, 39

*obtusum* Zone 14
Oil shale 40
Okeford Fitzpaine 75
OLDHAM, I.C.B. 65
*opalinum* Zone 18
*Ophiomorpha* 71
Osborne Beds 90
Osborne Member 108
Osmington Hill 47
Osmington Mills 40, 39
Osmington Oolite 39, 32, 38
''Osses 'eads' 50
OWEN, H.G. 74, 75
Oxford 38
Oxford Clay 30, 32, 122
*oxynotum* Zone 10, 14
Oyster Bay 109

Paddy's Gap 105
Palaeocene 94
Palaeogene 90
PALMER, C.P. 16
PALMER, J.T. 30
PARSONS, C.F. 20
Passage Beds 38, 75
'Patch reefs' 50
Pebble Bed 72
PENN, I.E. 26, 27, 28
Pennard Sands 16
Pentacrinite Bed 14
Periglacial Involutions 128
'Perna' Bed 70, 72
Phosphatic nodules 75
*pilula* Zone 88
'Pine Raft' 65
Pipeclay Beds 100, 102
'Planktonic Datum' 96
*Planolites* 83
Plateau Gravel 124, 125, 128
Pleistocene 124
Plenus Marls 86

Pliocene 126
POMEROL, C.H. 90, 94, 96
Pondfield Cove 50
Pondfield Member 47
Poole Harbour 23
Popple Bed 80
Portesham 39, 40, 50, 53, 60, 79
Portland Bill 12, 126
Portland Clay 48
Portland Group 30, 43, 56
Portland Harbour 32
Portland Limestone Formation 43, 53
Portland Limestone 47
Portland Raised Beach 130
Portland Sand 53
Portland Sand Formation 43, 47, 50
Portland Shell Bed 48
Portsdown 39, 61, 64, 67
Portsdown Anticline 121
Portsdown Borehole 122
Portsdown Swell 39
Portsmouth 95, 96
Potterne 53
Powerstock 26
Poxwell 50
Poxwell Circus 50
Poxwell Fault 121
Poxwell Pericline 114
Pre-Albian movement 114
'Prickle Bed' 48
Puckaster Cove 71
'Puffin Ledge' 48
PUGH, M.E. 130
Punfield 67
Punfield Cove 72, 75, 79
Punfield Marine Band 72
Purbeck Anticline 114
Purbeck Fold 50
Purbeck Group 7, 30, 53, 54, 62
Purbeck Marble 62
Purbeck Monocline 114

Ragstone Beds 53
Raised Beach-100 feet 125
Raised Beaches 125
Rampisham 79
Rams Down 75
*raricostatum* Zone 14
Reading Beds 90, 94
Recent 124
Red Band 14
Red Cliff 71, 72, 74
Red Nodule Beds 32
Redcliff Point 32
Redend Sandstone 100

Reeth Bay 71
REID, C. 125
*Rhaxella* 48
*Rhizocorallium* 8
Rhynchonella Marls 43
RIDD, M.E. 114
Ridgeway 64
Ridgeway Fault 114
Ringstead 32, 38, 39, 40
Ringstead Bay 40, 43, 47, 50
Ringstead Coral Bed 32, 38
Ringstead Waxy Clay 38
River Avon 30
River Char 13
River Hamble 122
River Solent 125, 130
'Roach' 50
Rodden Hive Point 25
Rotunda Nodule Bed 40, 43
Rotunda Nodules 50
Ryde 108
Rye Hill Farm 80
Rye Hill Sand 80


Sandown 71
Sandrock 71
Sandrock Series 70
Sandsfoot Castle 38
Sandsfoot Clay 32, 38, 39
Sandsfoot Grit 32, 38, 39
Saurian Bed 10
Scallop Beds 62
SCANES, J. 80
Scaphites Beds 70
*scissum* Zone 20
Scroff 25
Seatown 14
Seend 72
SELLWOOD, B.W. 7, 8, 14, 43
Selsey 102
Selsey Bill 100, 126
Selsey Formation 100
Selsey Raised Beach 130
SENIOR, J. 17
Serpulites 50
Shaftesbury 72
Shales with Beef 10, 13
Shanklin 71
SHEARMAN, D.J. 130
Sherborne 16, 18, 20, 22, 25, 26, 27
Sherborne Park 25
Shillingstone 79, 80
Shortlake 32
Shrimp Bed 48

Sicilian Terrace   128
Sleight Terrace   128
Slindon Raised Beach   128
SMALL, R.J.   130
SMITH, A.J.   91
Smithi Limestone   28
'Snuff Boxes'   20, 21
Soft Cockle Beds   56, 60
Solent   130
Solent Formation   105
Solifluction   125, 130
Southampton   95
Southampton Water   100, 121, 130
Southern Downs   75
Southwell   60
*spinatum* Zone   10
Spurious Chalk Rock   86
St Albans Head   40, 47, 48, 50
St Catherines Point   71
St Gabriels Water   14
St Helens Sands   108
Stalbridge   20
Starfish Bed   16
Steeple Ashton   32, 38
Sticklepath Fault Zone   121
STINTON, F.C.   93, 94, 100, 102
Stokeford Heath   128
Stonebarrow   74
Stonebarrow Cliff   13
Stonebarrow Hill   14, 16
Strahan, A.   64
Stromatolite   20, 53, 56, 60
Studland Bay   90, 94
Sturminster Newton   38
Submerged Forests   125
SURLYK, F.   32
Sutton Poyntz Pericline   114
Swanage   7, 64, 67
Swanage Bay   67, 72
Sway   128
'Swell'   48, 53, 54
Swindon   43, 50
SYKES, R.M.   32

Table Ledge   10
TAITT, A.H.   114
TALBOT, M.R.   32
TE PUNGA, M.T.   130
*tenuicostatum* Zone   17
Tertiary   90
Tethys   17
Thalassinoides   8, 83
Thorncombe Beacon   16
Thorncombe Sands   16, 17
*thouarense* Zone   18
THOUVENIN, J.   96

*Index*   145

Three Tiers   16
Tidmoor Point   16, 32
Tilly Whim Caves   50
Tisbury   64
Tisbury Glauconitic Member   53
Toarcian Stage   17, 18
Todber Freestone   38
TORRENS, H.   56
Totland Bay   108
TOWNSON, W.G.   43, 53, 54
Trace fossils   7
Trigonia hudlestoni Bed   32
Twinhoe Beds   28

Undercliff   75
Unio Beds   62
Upper Albian   75
Upper Building Stones   62
Upper Chalk   88
Upper Clay   95
Upper Cornbrash   27, 30
Upper Crioceras Beds   71
Upper Cypris Clays   62
Upper Fullers Earth   25, 26, 27, 28
Upper Greensand   74
Upper Gryphaea Beds   71
Upper Hamstead Beds   109
Upper Headon Beds   108
Upper Inferior Oolite   20
Upper Jurassic   30
Upper Lias   17
Upper Lobster Beds   70
Upper Lydite Bed   43, 50
Upper Rags   28, 30
Upwey   64, 67
Urchin Bed   71

Vale of Pewsey   38, 40, 53, 62, 75, 86, 90
Vale of Wardour   38, 40, 50, 53, 61, 62, 64, 67, 72, 75, 88
Valley Gravel   124, 125
Variscan Orogeny   112, 114, 121
Venus Bed   108
Verwood   95, 96
*Viviparus*   62
Viviparus Clays   64
Volcanic Ash   27

Walpen Clay and Sand   71
Walpen Undercliff   71
Wardour Member   50, 53
Wareham   67, 100, 102, 128
Warminster   80
Warminster Greensand   80

146  *The Hampshire Basin*

Watershoot Bay   71
Watton Cliff   25, 26
Wattonensis Beds   25, 26
Wealden   62, 64, 67
Wealden Marls   65
Wealden Shales   67, 72
West Basin   43, 48
West Bay   126
West Cliff   10
WEST, I.M.   56, 121, 130
West Sussex   125
West Weare Sandstones   47
Westbury   20, 23, 27, 32, 38
Westbury Iron Ore   32
Westbury Ironstone   38
Weymouth   26
Weymouth Anticline   50, 114
Whales Chine   71
*wheatleyensis* Zone   40
'Whitbed'   50
Whitchurch Sands   62, 67
White Band   110
Whitecliff Bay   94, 95, 96, 100, 102, 105, 108
WHITTAKER, A.   114, 121
WILLIAMS, R.B.G.   130
WILSON, R.C.L.   39
WILSON, V.   26
WIMBLEDON, W.A.   43, 48, 50

Wimborne   91, 94, 128
Wimborne Minster   114, 122
Wincanton   20
Winchester   67
Winchester Anticline   114, 122
Winspit Member   48, 50
Winterborne Fault   121
Winterborne Kingston Borehole   122
Wittering Formation   100
Wockley Micritic Member   53
Woking   91
WOODWARD, H.B.   39
Wool   128
WOOLRIDGE, S.W.   128
Wootton Creek   108
Worbarrow Bay   61, 64, 67, 72, 79
Worbarrow Tout   60
Worth Matravers   50
WRIGHT, C.A.   38, 95, 96, 102
WYATT, R.J.   26, 27, 28
Wytch Farm Boreholes   122

Yaverland Fort   72
Yeovil   18, 20, 22, 114
Yeovil Junction   20
Yeovil Sands   18, 20

ZIEGLER, B.   40
Zigzag Bed   25

Printed in the U.K. for HMSO.
Dd 717074 C200